ALL THUMBS

Guide to
VCRs

Gene B. Williams

Illustrations by Patie Kay

TAB | **TAB BOOKS**
Blue Ridge Summit, PA

FIRST EDITION
SECOND PRINTING

© 1993 by **TAB Books**.
TAB Books is a division of McGraw-Hill, Inc.

Library of Congress Cataloging-in-Publication Data

Williams, Gene B.
    VCRs/by Gene B. Williams.
       p.   cm.
    Includes index.
    ISBN 0-8306-4181-5
    1. Video tape recorders and recording.   I. Title.
TK6655.V5W555 1992
621.388'33--dc20   92-9121
                              CIP

Editorial Team: Roland S. Phelps, Acquisitions Editor
                 Melanie D. Brewer, Book Editor
Design Team: Jaclyn J. Boone, Designer
               Brian Allison, Associate Designer
Production Team: Katherine G. Brown, Director of Production
                   Wanda S. Ditch, Layout
                   Susan E. Hansford, Typesetter
                   Kelly S. Christman, Proofreader
                   Joan Wieland, Proofreader
Cover Design: Lori E. Schlosser
Cover Illustration: Denny Bond, East Petersburg, Pa.
Cartoon Caricature: Michael Malle, Pittsburgh, Pa.   ATS

The All Thumbs Guarantee

TAB Books/McGraw-Hill guarantees that you will be able to follow every step of each project in this book, from beginning to end, or you will receive your money back. If you are unable to follow the All Thumbs steps, return this book, your store receipt, and a brief explanation to:

All Thumbs
P.O. Box 581
Blue Ridge Summit, PA 17214-9998

About the Binding

This and every All Thumbs book has a special lay-flat binding. To take full advantage of this binding, open the book to any page and run your finger along the spine, pressing down as you do so; the book will stay open at the page you've selected.

The lay-flat binding is designed to withstand constant use. Unlike regular book bindings, the spine will not weaken or crack when you press down on the spine to keep the book open.

Dedication

To Tommy Kay who did more than he knows.

Acknowledgments

A number of people helped in getting this series started and this book put together. Most of all I would like to thank Patie Kay, my artist, "partner," and close friend. She not only created all of the art used in this book but spent countless hours reading and critiquing the material. Without her this book wouldn't be nearly as good, and certainly wouldn't have been as much fun.

Melanie Brewer of TAB Books served as the book editor. Her dedication to the job and her patience as we struggled to get the format just right did much toward making the final book possible.

Roland Phelps, also of TAB, oversaw the project and helped to smooth the many bumps and turns that had to be made while getting the format to work.

And of course there is Cindy, my wife, and Danny, my son, who patiently tolerated all the long and odd hours.

Contents

Preface

A collection of books about do-it-yourself home repair and improvement, the All Thumbs series was created not for the skilled jack-of-all-trades, but for the average homeowner. If your familiarity with the various systems in the home is minimal, or your budget doesn't keep pace with today's climbing costs, this series is tailor made for you.

More than two-thirds of all homes in America have at least one VCR, and the number continues to grow. Many won't believe that you can take care of at least 90% of any malfunctions and nearly 100% of the maintenance yourself. Follow the maintenance steps provided in this book and the number of malfunctions will be greatly reduced. You'll face fewer repairs and less serious repairs (many of which you will be able to handle yourself). Also, your VCR will last longer. The All Thumbs series saves you time and money by showing you how to make most common repairs yourself.

The guides cover topics such as home wiring; plumbing; painting, stenciling, and wallpapering; repairing major appliances; and home energy savings, to name a few. Copiously illustrated, each book details the procedures in an easy-to-follow, step- by-step format, making many repairs and home improvements well within the ability of nearly any homeowner.

Introduction

Designed for the novice do-it-yourselfer, this guide walks you through the troubleshooting necessary to properly maintain your VCR and the cleaning process that often solves the majority of VCR malfunctions. When it comes time for repair, the simple step-by-step instructions in this book will make you successful the very first time you try.

Chapter 1 explains the importance of safety. All the tools you'll need and how to use them are described in chapter 2. Chapter 3 explains the controls common to most VCRs and chapter 4 walks you through the basics of how a VCR works. With this essential but simple background, you're ready to open the VCR. This is easier than you might suspect. Chapter 5 tells you exactly how to do it and chapter 6 explains how to clean your VCR to keep it in top condition.

Chapter 7 details some of the simple checks and repairs you can make, and the tapes themselves are covered in chapter 8. Make the connections that will turn individual units into the ultimate system with the guidance of chapter 9.

The purpose of this book is to help you understand that you have fewer limitations than you suspected. You'll be surprised at just how much you can do. Follow the directions in this book and your VCR will work better and last longer.

When you open the case of your VCR and take some troubleshooting steps, be sure to write down what you do. Even if you are unable to fix the problem, you can save yourself money by being able to help the technician spot the problem faster.

Also with these notes, you are less likely to pay for unnecessary work done to your VCR. Running across dishonest technicians is relatively rare, but it can happen. By being aware, and having your own preliminary notes, you can eliminate this almost completely.

If your VCR is still under warranty, it's generally best to let the authorized service center take care of any problems. The warranty generally covers everything except cleaning. In many cases if you open the cabinet while the VCR is under warranty, you could void that warranty; however, not much is likely to go wrong during the warranty period.

Often you'll get a list of authorized centers and sometimes phone numbers to call when you buy your VCR. In some cases the place where you purchased the VCR will provide any warranty service. If the VCR you want to buy has no local service centers, you'll have to box and ship the unit any time you need to have it repaired.

CHAPTER ONE

Safety

Safety is essential for you to know and practice. It is important for everyone, especially for beginners who might not know all of the potential dangers. Most of the circuits in the VCR present no danger to you. However, the power coming out of the wall outlet and going into the power supply can measure 120 lethal volts.

Getting a jolt from the incoming 120 volts ac (120 Vac) is more than just unpleasant; it can be fatal. Studies have shown that it takes very little current to kill. Even a small amount of current can paralyze your muscles and you won't be able to let go. Just a fraction more and your heart muscle can become paralyzed.

There is almost nothing mechanical in the VCR that can harm you. About the most that can happen is you'll get a pinched finger. There is however a greater threat to the VCR. Some of the parts can be easily damaged. Touching the wrong place can make a short circuit and cause a critical circuit to go up in smoke. Moving parts can get broken, bent, damaged, or knocked out of alignment. If any of these things happen, the VCR will not work correctly. If something goes wrong, more damage could result because all the parts (including the tape itself) work together.

Now that you are sufficiently frightened, relax. As long as you are careful, take your time, and follow rules of safety and common sense, you won't get into trouble. Learn where the dangerous spots are, stay away from them, and you'll have no problems.

Before beginning any work inside the VCR, remove all jewelry, especially jewelry that dangles. A lot of jewelry is made of metal, which is electrically conductive. You might short a circuit or get shocked if a ring or necklace comes in contact with something it shouldn't.

Every time you work on your VCR, exercise the following steps.

Step 1-1. Unplugging your unit.

Unplug the VCR before opening the cabinet. Doing this might mean having to reset the clock and other controls, but you can't possibly get shocked, and you won't accidentally cause a short circuit that can damage the VCR. The correct way to unplug any appliance or device is to grasp the plug. Never pull on the cord, which can cause it to pull loose from the plug.

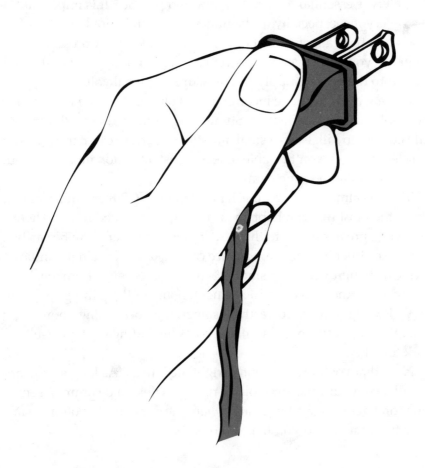

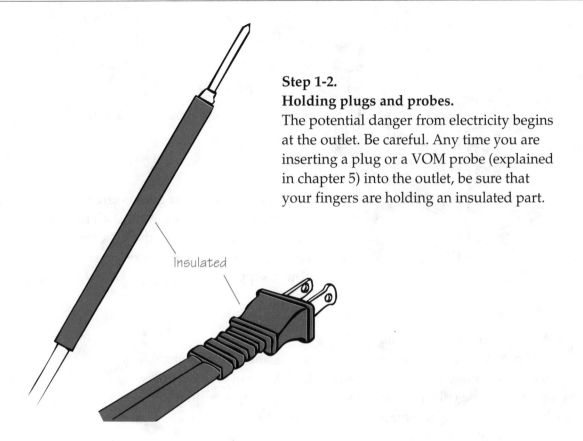

Step 1-2.
Holding plugs and probes.

The potential danger from electricity begins at the outlet. Be careful. Any time you are inserting a plug or a VOM probe (explained in chapter 5) into the outlet, be sure that your fingers are holding an insulated part.

Insulated

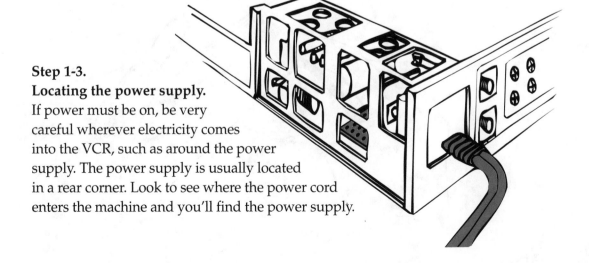

Step 1-3.
Locating the power supply.

If power must be on, be very careful wherever electricity comes into the VCR, such as around the power supply. The power supply is usually located in a rear corner. Look to see where the power cord enters the machine and you'll find the power supply.

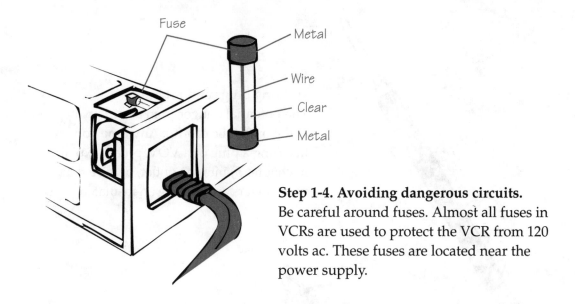

Step 1-4. Avoiding dangerous circuits.
Be careful around fuses. Almost all fuses in
VCRs are used to protect the VCR from 120
volts ac. These fuses are located near the
power supply.

Step. 1-5. Probing inside your VCR.
Metal probes and tools can cause short circuits and other damage.
There is no danger if you've unplugged the VCR, but at other times
power will have to be present. At such times, paying attention to
what you are doing becomes more important than ever.
The chassis of your VCR might be metal.
When you probe, be careful not to
touch a test point and the metal
chassis at the same time or you
might get shocked.

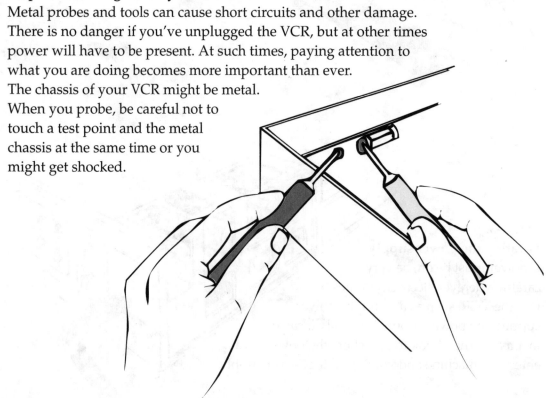

Step 1-6.
Avoiding contact with the video head assembly.
When working inside the VCR, move slowly and carefully. It's easy to cause damage to the delicate parts. Be especially careful around the video head assembly, and never touch it with your bare fingers. Touching the assembly in this manner can cause permanent damage; so can excessive pressure. Alignment of the assembly is critical. The heads mounted in the assembly are even more delicate.

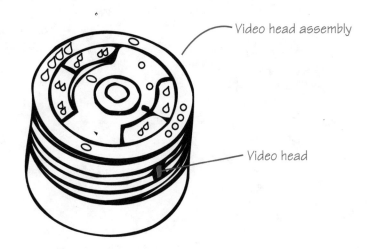

Video head assembly

Video head

CHAPTER TWO

Tools & Materials

P robably the tools you'll need to troubleshoot and repair your VCR can be found around the house. None of the tools and materials are costly. The most expensive is the optional VOM; a unit costing about $10 to $20 will be sufficient. Avoid buying poor-quality tools. A cheap screwdriver might do the job just as well as an expensive one; however, it also can cause you considerable trouble. There are reasons why a cheap tool is cheap. The materials used are of lower standards. The metal can bend and break, or the coating can flake off and fall inside the machine where it can do considerable damage.

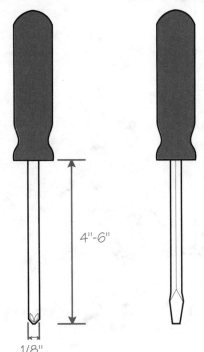

Screwdrivers Invariably, the cabinet of the VCR will be held by Phillips-type screws, as are most internal shields and many other parts. The video cassette itself uses this kind of screw to hold it together. The head of a Phillips screwdriver is in the shape of an x. Not needed as often but still handy is the standard blade-type screwdriver. In both cases, get a tool with a medium-sized head (about 1/4 inch) and one with a small-sized head (about 1/8 inch). The shaft length isn't important, but most people find a length of 4 to 6 inches to be the most convenient and versatile. Be sure that the screwdrivers are not magnetized.

The screwdriver should fit the screw, both for the length of the slot and for the width. The same is true of the Phillips screw and screwdriver. The head should fit the slot.

Needle-nose pliers   A pair of needle-nose pliers is good for light gripping, parts retrieval, and various other tasks. A standard pair of pliers also might come in handy. In both cases, be sure that the handles are insulated.

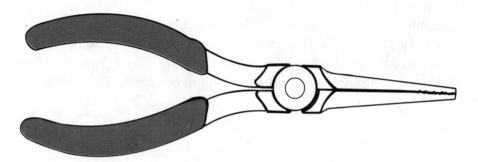

Isopropyl alcohol   This is the standard cleaning fluid. It's inexpensive and relatively safe to use. For a VCR, be sure to get technical grade isopropyl alcohol with a purity of at least 95 percent. *Caution!* Do not use alcohol on any rubber parts.

Special cleaning fluids are more expensive than alcohol; however, they are often better cleaners. Get a cleaner that leaves no residue and is safe for rubber parts. Usually the label will tell you.

Cleaning Swabs   The swabs used must be of the appropriate material. Those with threads, like cotton swabs, are just barely tolerable for general inside cleaning. They should never be used on the video head assembly. Instead, use the foam swabs found in most electronics and video supply stores.

Optical-grade chamois is suitable for cleaning the video head assembly. These can be purchased at optical stores and at some camera stores. Call around first; they're not always easy to find. You should not use automotive chamois because they might be chemically treated.

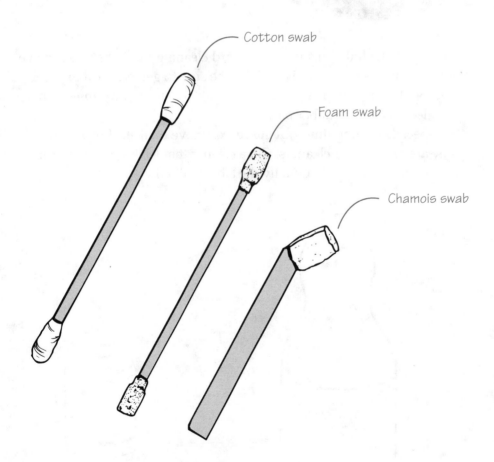

Cotton swab

Foam swab

Chamois swab

Cleaning Cartridge   Manual cleaning is always preferable, when possible. Sometimes, such as with some camcorders, you cannot get at some of the parts safely. At such times you'll need a cleaning cartridge. Use only the "wet" variety.

Volt-ohmmeter   A volt-ohmmeter, also called a VOM or multimeter, is used to measure voltage and resistance. It is an invaluable tool to have around the house because you can make power tests, continuity tests, and other tests. Most of the tests you'll be making will not require a high degree of accuracy, so an inexpensive VOM works fine. (See the Appendix.)

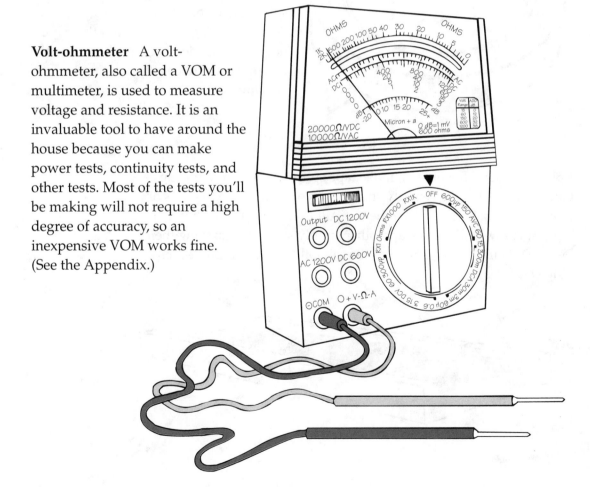

This final section deals with various tools you may or may not need. You don't have to buy any of these tools until the need arises.

Many of the things you'll need can be found around the house. A muffin tin or paper cups can be used to hold removed screws or other parts. A pair of scissors is needed for many tape repairs.

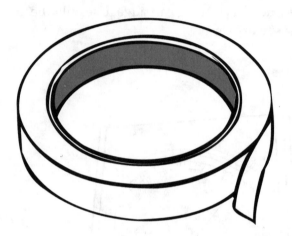

Splicing tape   When tape repairs are needed, buy 1/2 inch splicing tape meant specifically for video tape. Unfortunately this tape isn't always easy to find. If you can't find it, a suitable substitute is regular adhesive tape (e.g., Scotch or "Magic Mend,") but these are not recommended because they do not match the width as efficiently as the splicing tape.

Splicing block   To help you make tape splices, invest in a splicing block. It holds the tape exactly straight so you don't have to guess or fiddle. Splicing blocks used to be difficult to find for 1/2-inch video tape, but now Radio Shack and other supply stores are carrying them. Be sure that the one you buy has the capability of holding 1/2-inch video tape.

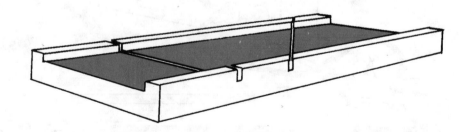

The reason that you'll almost never hear about video tape splicing (other than in this book) is because it's generally a bad idea and brings a danger to the VCR. Each time that splice goes across the video head assembly, it might cause damage. To reduce this risk, a repaired tape should be copied (dubbed), and the repaired original should be set aside to be used only as an emergency backup. To make this copy, you will need a pair of dubbing cables. These are usually about 3 feet long and have an RCA (phono) plug on each end.

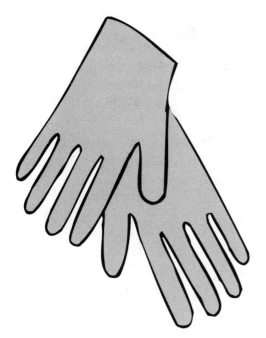

Lint-free gloves   Your bare fingers should never touch the video tape. If you intend to make any repairs or do any splicing, get some lint-free gloves. These gloves are inexpensive and can be found at almost any photo supply store.

CHAPTER THREE

Controls &
Connectors

O ver the years, the most common complaints I've heard are that the controls are too difficult and too confusing, and that the buttons are too small. Part of the problem is that there is little or no uniformity between manufacturers, and that even with the same manufacturer there might be significant differences between models.

An entire book could be written about the controls and the variations. All the possibilities still would not be covered. Don't be surprised if the illustrations that follow aren't exactly the same as the controls on your specific machine. Your best guide is the instruction manual for your specific make and model.

Learning at least the basic controls and how they function is one of the first steps in both understanding your VCR and in troubleshooting it.

VCRs are both different and much the same from model to model. Many of the controls, such as Play and Stop, are self explanatory. The Fast Forward and Rewind buttons usually have two functions. If the machine is in the Stop mode, these controls move the tape quickly forward or backward. If the machine is in the Play mode, these buttons perform a Search or Scan. The Record feature might require that you press Record and Play simultaneously, or it might be a single-button function (in which case the button is almost always red).

Front panel controls

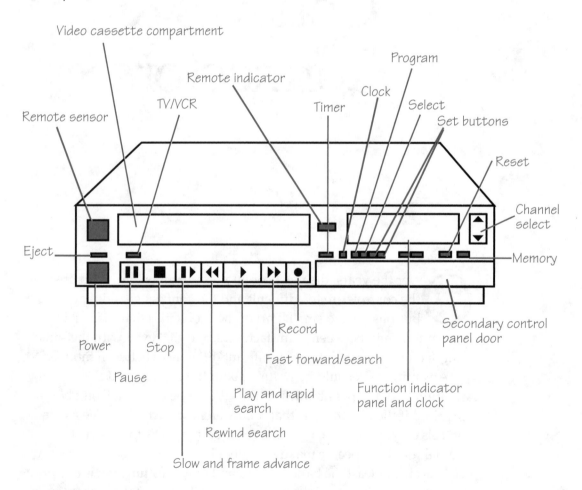

Most VCRs also have a Timed Recording button (often located by itself on the right side). It is usually labeled with three letters, such as OTR (One-Touch Recording), IRT (Instant Recording Timer), or XPR (Express Recording). This button allows you to begin recording immediately without touching the Record button and to continue recording, usually in increments of 15 or 30 minutes. When the time has elapsed, the recording automatically stops. The Pause button brakes the tape movement, allowing you to hold the picture still. If the unit has a Frame Advance to view the recording one frame at a time, it is usually located on the remote control rather than on the unit.

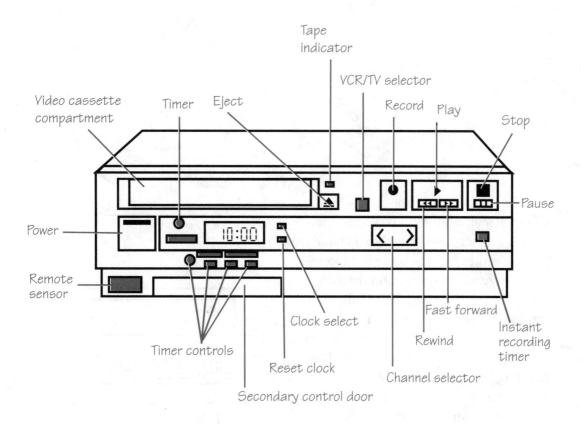

VCR front

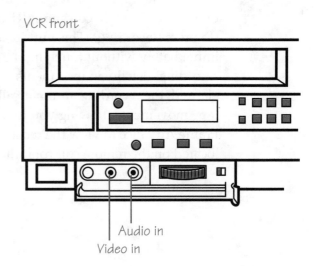

Front panel connections for audio/video in/out used to be fairly standard to make VCR-to-VCR copying easier. For a while this feature disappeared, but it is now coming back again.

Audio in

Video in

TV/VCR switch

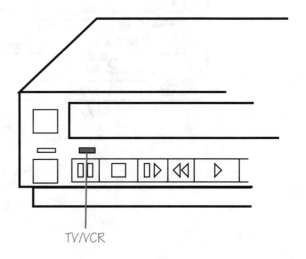

Also on the front panel, the TV/VCR switch allows you to use the VCR as a tuner. This capability is handy because sometimes the tuner section in the VCR gets better reception than the one in the television set. It also allows you to watch one channel on the television while recording on a different channel. (In this case, set up the VCR, start recording, press the TV/VCR switch, and retune the television to whatever channel you wish to view.)

TV/VCR

Auxiliary controls are located behind a door or panel, usually at the bottom of the front panel. Inside are the controls for other functions, such as programming. The exact procedure for using these controls varies widely from unit to unit. Setting the clock is usually much like setting any digital clock, and it must be done before any timed recordings can be made. To program those timed recordings, you must set a beginning time and an ending time, plus the day and often the choice of week. You also need to select the channel. Often the day, week, and channel will remain at the present settings unless you purposely change them.

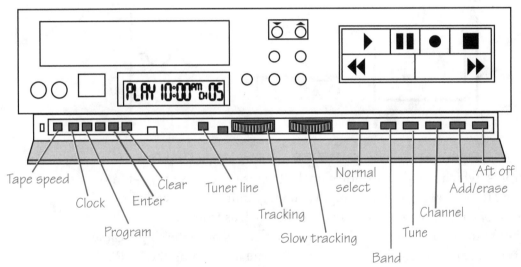

Auxiliary controls

The tracking control adjusts the video head so that the path of the tape and the timing of the video heads match what is needed by the recording.

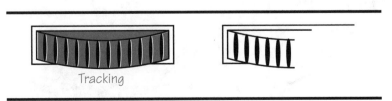

Tracking

At least one control, the selector for the output channel of the VCR, is located in the back of the VCR. With rare exception, your choice will be limited to either channel 3 or 4. The channel you select depends on your location. If there is a strong local channel 3, but no channel 4, the output selector should be set to 4. Likewise, if you get good reception from the TV station on channel 4, the switch should be set to 3. If you get both, set the switch to whichever is the weakest. The purpose is to reduce interference.

The back of the VCR

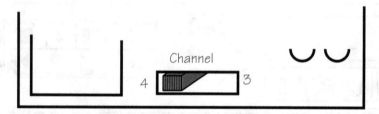

Also located in the back of the VCR are panel signal inputs and outputs, which are how outside signals (television, cable, etc.) get into the VCR for viewing or recording. With most units, you have your choice of connecting the incoming signals, such as from an antenna, either to the two screw-type (300-ohm) connectors or to the single cable (F-type) composite connector. The same is true for sending the VCR's signal to the television set. Chapter 9 discusses connections and gives you more information on the wires, cables, and connectors needed.

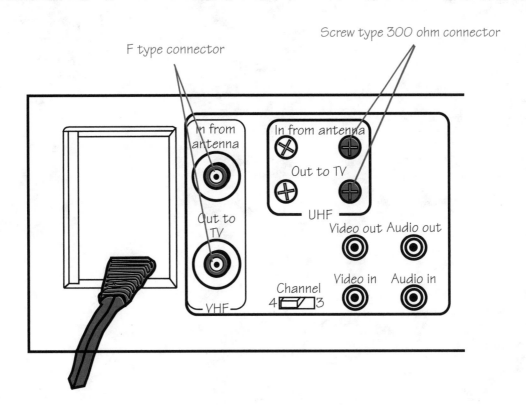

F type connector

Screw type 300 ohm connector

In from antenna

In from antenna

Out to TV

UHF

Out to TV

Video out Audio out

Video in Audio in

Channel
4 ☐ 3

VHF

The four audio/video inputs/outputs are valuable features that many people never use. (Chapter 9 shows you how.) The most common use is when two VCRs are hooked together to make dubs. In some cases these connectors are the preferred way to connect the VCR to a television. The audio out also can be used to send the sound through your home stereo—a handy tip if your VCR is monaural and you want to simulate stereo, or if you just want the better sound of the larger speakers.

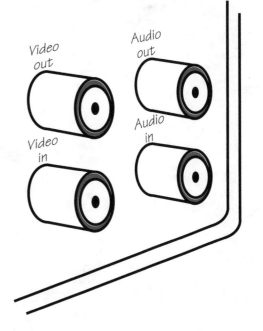

Video out

Audio out

Video in

Audio in

CHAPTER FOUR

How Your VCR Works

Imagine trying to fix a car without having any idea about what makes it work. You generally don't need this understanding if you will be merely operating the device or machine, but the knowledge is essential when it comes to troubleshooting and repairing. Your goal isn't to learn how to design and engineer a VCR. All you need to know are the basics of how it operates.

For example, the clock/timer of the VCR is operated by one or more electronic chips (integrated circuits) and related circuitry. You do not need to know how to design such a circuit to be able to push the buttons that set this clock/timer. Likewise, you don't need to know how to design a power supply to understand what it does and how to make the simple tests to see if it is working.

Learning how a VCR works isn't difficult. Your greatest enemy is "high-tech intimidation." Being cautious and recognizing your honest limitations is wise. Being intimidated to the point that you feel totally incompetent isn't necessary.

Up until now you've probably just shoved a cassette into the VCR, pressed Play, watched the movie, and never thought much about how it all happened. Because you bought this book, you must have at least some interest in learning how things work.

In chapter 6 you'll learn how to open the VCR so you can see it in operation if you wish. Seeing the VCR in operation while it is in good condition will give you valuable knowledge and clues for when things aren't working.

The major internal parts of the VCR include the erase head, video head assembly, linear audio head, rollers, capstans, guide pins, and other parts. These parts are illustrated in chapter 6.

When you insert a cassette into a VCR, a switch is tripped causing the VCR to "grab" the cassette, pull it inside, and lower it into position. As this happens, a pin slides into a hole in the bottom of the cassette, releasing the reels. Another pin releases the door catch on the cassette so the door can be opened. Eject does the same thing but in reverse.

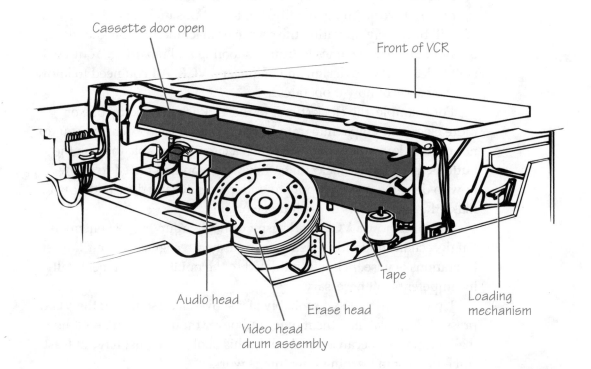

Play or Record causes various mechanisms to pull the tape from the cassette and place it correctly in the tape path. Pressing Stop does the same thing but in reverse, with the tape being placed back into the cassette. The door will stay open until the cassette is ejected from the VCR.

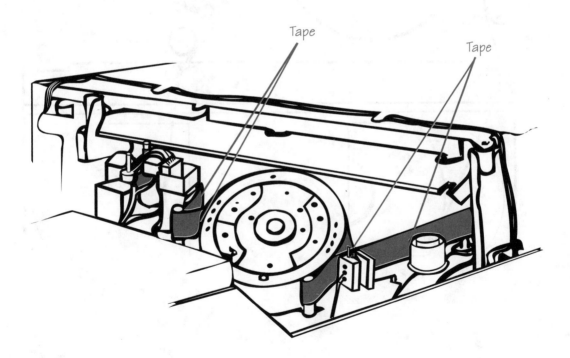

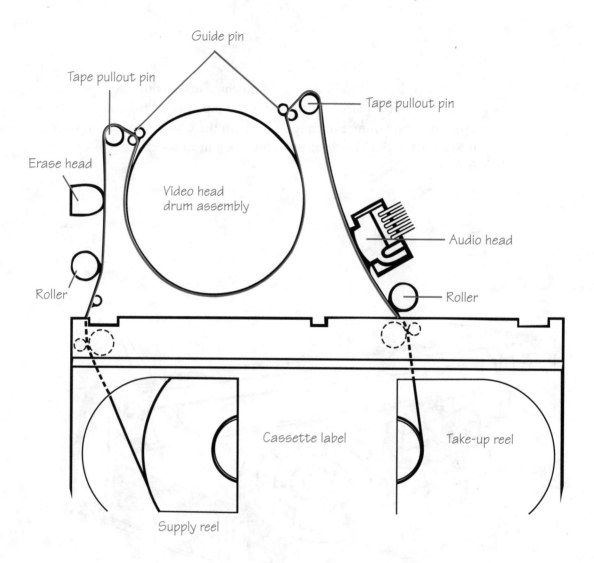

Guide pin

Tape pullout pin

Tape pullout pin

Erase head

Video head
drum assembly

Audio head

Roller

Roller

Cassette label

Take-up reel

Supply reel

When you get through chapter 6, compare your own machine while playing to these illustrations. Use caution because the cover will be off while power is being applied.

Tape recording, whether it be with an audio cassette or in a VCR, uses magnetism. Recording is done with a small electromagnet called the record/playback head (or heads).

The tape consists of a plastic base on which there is a coating of magnetic particles. These particles are extremely small. Great care must be taken during manufacture that they are evenly distributed on the base and that the binder used will keep them there. The first ensures a better recording. The second reduces *dropouts* and also reduces damage to the machine. A dropout is caused when flakes of the coating fall off the tape. Any magnetic field recorded there falls off with the flake. What you see on the screen as a result is a small white speck. If the problem is bad enough, the image on the TV might seem to almost sparkle. This indicates that the tape is either of poor quality or is old

Magnetic oxide coating

Plastic base

and worn. (The same symptom could mean that the video heads are in bad need of a cleaning.) No matter how well the tape is made, however, a little of that coating is left behind on the parts of the VCR each time; a well-made tape will leave minimal deposits. These deposits are one reason why regular cleaning is necessary.

Look at the illustration showing the recording of an audio signal. The sound waves originating from the speaker hit the microphone, which causes the microphone to produce an electrical current that is "in step" with the sound itself. This current is then amplified so that it can drive an electromagnet. An electromagnet is a device that converts electricity into magnetic fields. These fields fluctuate with various signals, which affects magnetic particles on the tape. The magnetic particles store (record) the varying fields.

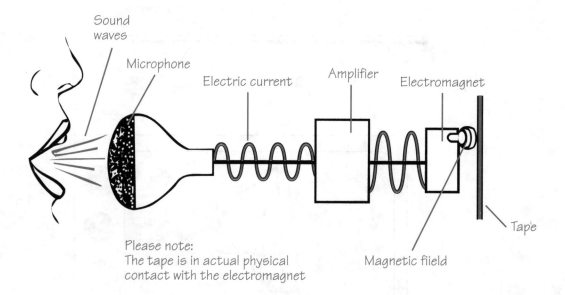

Sound waves
Microphone
Electric current
Amplifier
Electromagnet
Tape
Please note:
The tape is in actual physical contact with the electromagnet
Magnetic fiield

In playback, the stored fields recorded on the tape move over the playback electromagnet. The magnetic fields now cause an electrical current that is the same as the one that originally caused the recorded fields. (The tape is nearly in actual physical contact with the electromagnet.)

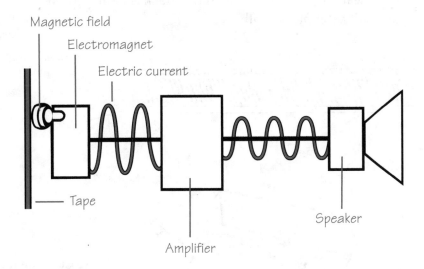

Magnetic field
Electromagnet
Electric current
Tape
Amplifier
Speaker

The high complexity and frequency range of the video signal requires extremely high fidelity, which requires a high tape speed. In the home VCR high fidelity is accomplished by moving the tape forward slowly and spinning the video head quickly. The head is tilted so that the tape crosses it at an angle. The top half of the video head drum spins very fast in a clockwise direction while the bottom half remains stationary. This motion causes the video signal to be recorded and played back as diagonal stripes.

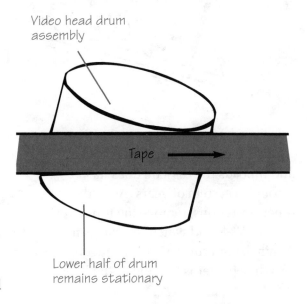

Video head drum assembly
Tape
Lower half of drum remains stationary

To keep everything working properly, a sync or cue track is necessary. The purpose of this track is to keep the tape and the VCR, and the VCR and television, in step with each other. The audio track is a linear strip at the top of the tape (with the stereo audio track, if there is one, recorded as diagonals much the same as the video). The sync track is a linear strip at the bottom. If this track is damaged, the VCR will play badly, if it plays at all.

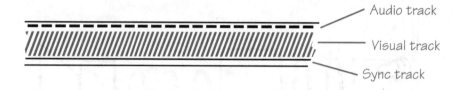

Audio track

Visual track

Sync track

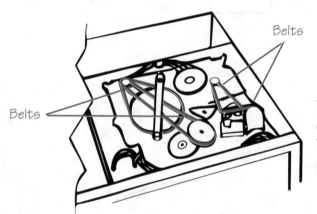

Belts

Belts

Most of the motors are located beneath the VCR. Often they are connected to the devices they drive by a system of pulleys and belts.

Sometimes the motor that loads and unloads the cassette does its job through a system of gears. As with other operations, remove the top cover of your VCR and watch this happen. Again, be careful! The cover must be off while power is flowing in the VCR.

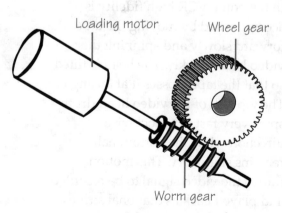

Loading motor

Wheel gear

Worm gear

CHAPTER FIVE

Basic Troubleshooting

Troubleshooting is a step-by-step process of elimination. Once you've eliminated all the places where the problem is not, you find out where it is. Some examples of common VCR malfunctions are given in this chapter. At times the procedures for solving common problems require tests inside the VCR with the power on. *Be Careful!* (Refer to chapter 1.) If you don't feel competent to do something that even *might* be dangerous to you or the VCR, don't do it! Call a professional technician.

The process always begins with the simple and progresses to the more complex. Start by looking for the obvious. Where you begin is determined by the general symptoms. For example, if the VCR is completely dead, trying a different cassette in the machine won't help. Likewise, if you obviously have power to the VCR, checking the house fuses or circuit breakers is a waste of time.

Once you've determined what is going wrong, you can begin to find out why the VCR is malfunctioning and what to do about it. A troubleshooting list, such as the one in the owner's manual or the one at the beginning of this chapter, can help.

The VCR system consists of three parts: the VCR itself, the tapes it plays or records, and the television set. Sometimes what seems to be a serious problem with the VCR or TV is actually nothing more than a bad tape. Often in the following steps, you'll be advised to try another cassette that you know is good. This same "test tape" can be used to make adjustments and is a valuable tool for keeping track of how well the VCR is operating.

Tools & Materials

- ❑ Voltohm-milliammeter (VOM)
- ❑ Cassette tape
- ❑ Cloth
- ❑ Cleaning fluid

TROUBLESHOOTING GUIDE

Problem	Probable cause	Solutions
No power	Wall outlet dead	Check outlet.
	Fuse or circuit breaker blown	Replace blown fuse.
	Power supply bad	
	Short circuit somewhere in VCR	
Fails to play	No tape	Insert cassette.
	Cassette has played to the end	Rewind cassette.
	Bad play switch	
	Bad power supply	
Fails to record	No tape	Insert cassette.
	Cassette has played or recorded to the end	Rewind cassette.
	Record-protect tab has been removed on cassette	Seal the area.
	Improper control operation	Study owners manual.

TROUBLESHOOTING GUIDE CONTINUED

Problem	Probable cause	Solutions
Poor-quality picture	Worn or damaged tape	Replace the tape.
	Dirty machine	Clean the VCR.
	Bad cables or connectors	Clean the cables.
	Tracking is maladjusted	Adjust tracking.
Picture bends at top	Worn or damaged sync track	Replace tape.
Mangled tapes	Poor quality tape	Replace the tape.
	Dirty machine	Clean the VCR.
	Worn idler tire	
	T-160 tape is being used	Use standard tapes.
Jumbled picture	Damaged tape	Change tapes.
	Tracking problems	Adjust tracking.
	Dirty VCR.	Clean the VCR.
Dropout	A tape with particles flaking off (an indication that tape is old or worn)	Change tapes.
	Dirty VCR.	Clean the VCR.
Flagging (usually occurring at the top)	Damaged sync track	Change tapes.
Jumping picture	Damaged tape	Change tapes.
	Damaged sync track	Change tapes.
	Dirty VCR	Clean VCR.
	Overpowering copyguard	Change tapes.

Step 5-1. Making a test tape.

Making a test tape is easy. The total cost involved is the price of a cassette. Make a recording of anything you wish while the VCR is in good operating condition. Make recordings at each of the three speeds. You also can repeat the sections by recording for 10 minutes on the fastest speed (2-hour mode), 10 minutes on the middle speed (4-hour mode) and 10 minutes on the slowest speed (6-hour mode), then repeat the sequence throughout the tape. If a part of the tape gets damaged, you'll be able to remove the damaged section and still have a valid test tape (see chapter 8). Label the tape "Test" or something similar, then set it aside until it is needed.

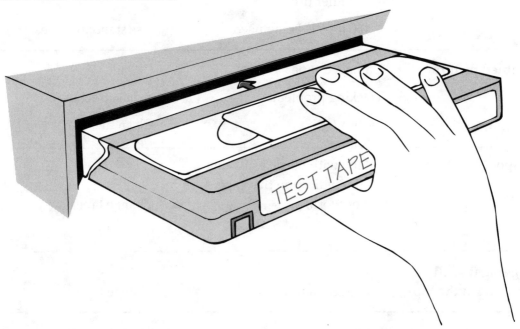

Step 5-2. Checking the obvious when nothing happens.

Troubleshooting begins by checking the obvious. If nothing at all is happening, look to see if the unit is plugged in. If it is plugged in, is it turned on? Another possibility is that the outlet is dead. Test the outlet by plugging something such as a lamp into it .

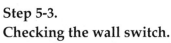

Step 5-3. Checking the wall switch.

If the outlet is dead, first look to see if the outlet is operated by a wall switch. Some homes have switches so that plug-in lamps can be turned on from a switch.

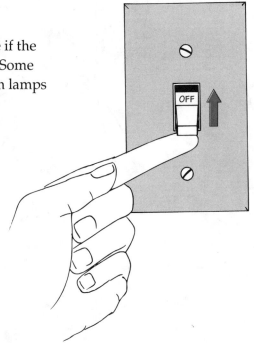

Step 5-4. Using a VOM.

For a more accurate test for power, a VOM (voltohm-milliammeter) can be used. Set the dial to read in the 120 volts ac range. Insert the probes into the outlet slots (being careful to hold the probes only by the insulation). Because the wall outlet has ac voltage, it doesn't matter which probe goes in which slot.

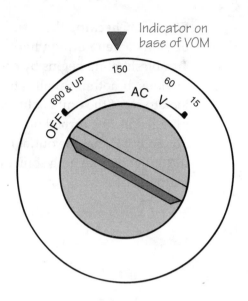

Indicator on base of VOM

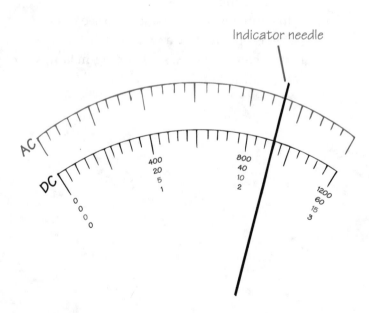

Indicator needle

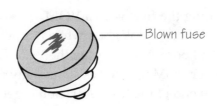
Blown fuse

Step 5-5.
Checking the
fuse or circuit breaker.

Usually you can see if the fuse wire is melted by looking through the "window" of the fuse. A popped circuit breaker might not be quite so easy to see. Normally the lever will be slightly back, but sometimes so slight that it is difficult to see. When in doubt, flip the breaker completely off and back on again.

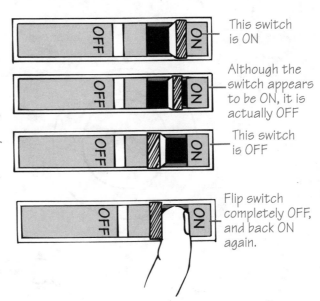

This switch is ON

Although the switch appears to be ON, it is actually OFF

This switch is OFF

Flip switch completely OFF, and back ON again.

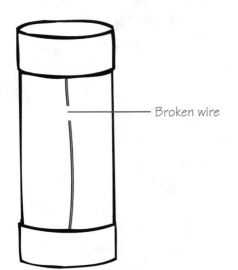
Broken wire

Step 5-6.
Changing the VCR fuse.

If the outlet is good but the VCR still won't "power up," the fuse in the VCR might be blown. You will have to open the cabinet to get at the fuse (see chapter 5). As with the household fuse, you can usually just look through the glass to see if the fuse wire inside is melted.

Step 5-7. Testing the fuse with a VOM.

Sometimes a fuse will look good when it is actually bad. If you suspect this, shut off all power and unplug the VCR. Remove the fuse from its holder. Set your VOM to read resistance; any range will do. Touch the probes to each end of the fuse. You should get a reading of nearly zero ohms. If the fuse shows a high resistance, it has "blown" and must be replaced.

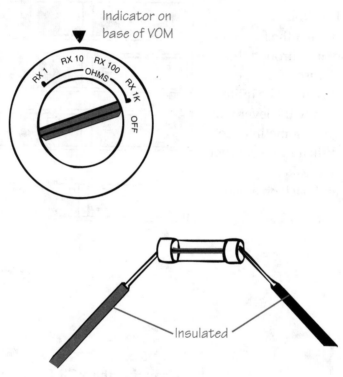

Indicator on base of VOM

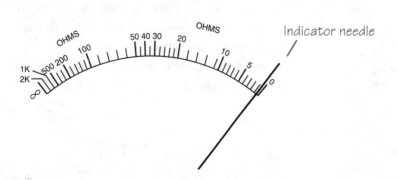

Insulated

Indicator needle

Step 5-8. Testing for power going into the power supply.

The final VOM tests involve the power supply. Be very careful doing these tests. The unit has to be plugged in, with the VCR power on. First, set the dial of the VOM to test for ac in the 120-volt range. Carefully touch the two probes to the spots where the incoming power cord is soldered. You can find these spots by simply looking at where the power cord enters the machine and following the cord inside. If you don't read approximately 120 volts ac, no power is getting to the unit. Either the outlet is dead, the cord is no good, or the soldered joints are bad.

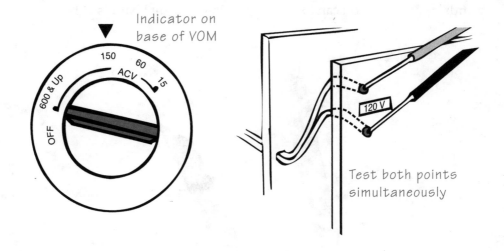

Indicator on base of VOM

Test both points simultaneously

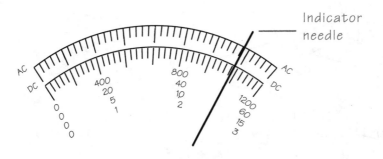

Indicator needle

Step 5-9. Testing for power coming out of the power supply.
If power is getting to the power supply, is it coming out? How you find out depends on how your VCR is built. Usually you only have to find where the wires come out of the power supply and look where they attach to a circuit board. Set the VOM to read dc voltage in a range above 12 volts (but not too far above). Basically, if power is coming in to the power supply but none is coming out you know that the problem is in the power supply. If power is coming out, and the VCR isn't working, the problem is somewhere in the VCR circuits. You might be able to make some tests by locating other test points on the circuit boards. Sometimes you can to determine which circuit board is at fault.

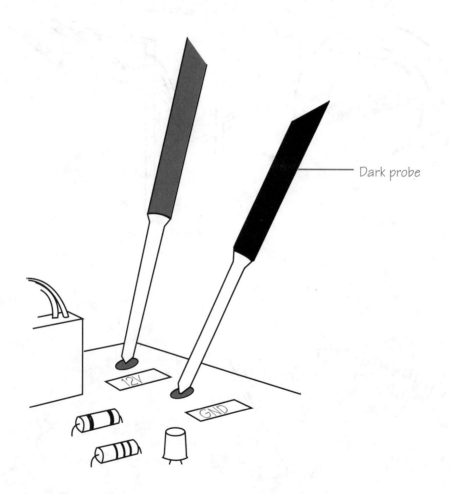

Dark probe

Step 5-10. Making sure the cassette is loaded correctly.

If you are getting power to the unit but it refuses to play, make sure you have a cassette in the VCR. Most VCRs have a light or some other indicator to show that a cassette is in. The cassette must be inserted with the label-side up; the door covering the tape must go in first. If the cassette is inserted in any other way, the VCR will not be able to load it, even if the VCR senses the cassette being inserted.

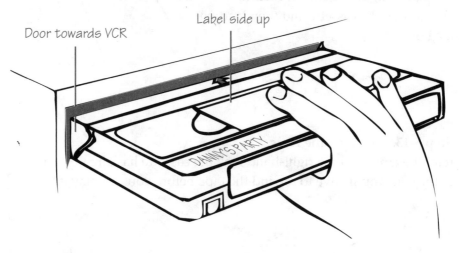

Step 5-11. Examining the cassette for damage.

Check the cassette itself. Is it of the right format? (A Beta cassette will not load into a VHS machine and vice versa.) Then visually examine the cassette for signs of warping or other damage. Set that cassette aside and try your test tape. If the second cassette loads and the other one will not, the problem is not with the VCR but with that first cassette.

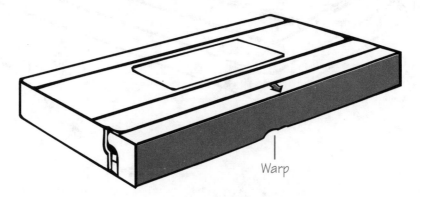

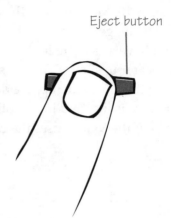

Eject button

Step 5-12.

Ejecting and reinserting the cassette.
It's possible that the cassette loaded, but
it didn't load flat. If it's "kinked" on
the platform, the VCR won't work.
Try ejecting the cassette and then loading
it again.

Step 5-13. Checking the reels.
If all the tape is on the right-hand reel, the cassette has already played
to the end. You'll have to rewind the tape before you can play it.

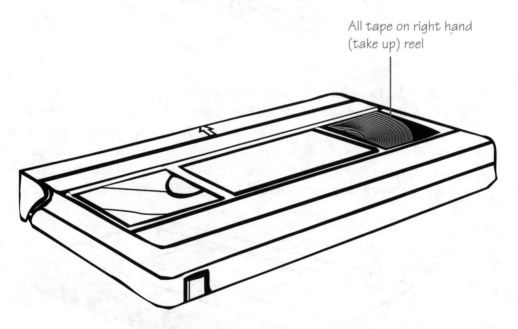

All tape on right hand
(take up) reel

Step 5-14. Examining the tape inside the cassette.
Press the lock button, located on the right-hand side just behind the door, and raise the door. Look at the tape. Does it show damage? If it does, you've located the problem. The damage doesn't have to be extensive to cause trouble. For example, if the damage is to the sync track (at the bottom of the tape), the VCR will have a difficult or impossible time trying to play the tape.

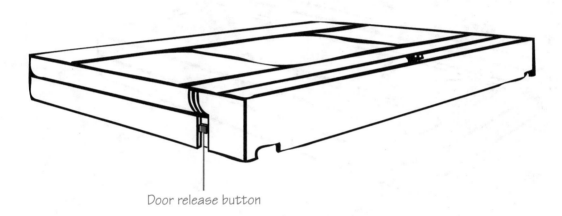

Door release button

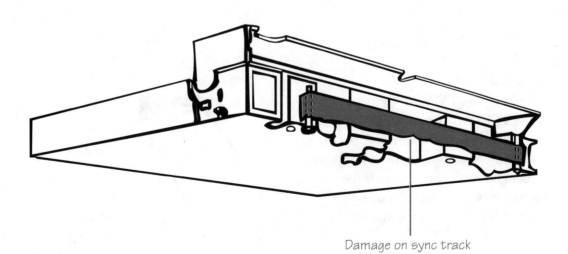

Damage on sync track

Step 5-15. Checking the record-protect tab.

If the tape will play but it refuses to record, eject the cassette and look at the back edge. The cassette has a record-protect tab on the left side. If this tab is in place, you can record on the tape (which means you should break this tab off to protect any valued recording). You will not be able to record if the tab is missing. If you are certain you wish to record on this cassette, you can cover the hole with a strip of adhesive tape.

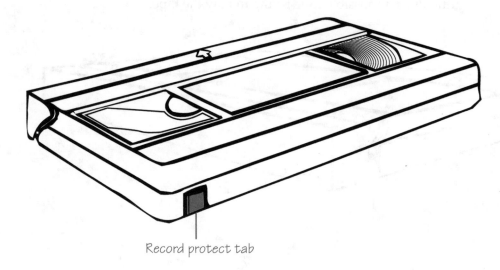

Record protect tab

Step 5-16.
Checking the TV/VCR switch.

Most VCRs will automatically change the TV/VCR switch when "Play" is started. To be sure, find this control on the front panel and press the button. If the image on the television doesn't change, it's not getting a signal from the VCR.

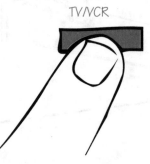

TV/VCR

Channel

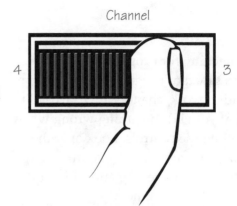

Step 5-17.
Checking the channel selector.
On the back of the VCR is a channel selector switch that allows you to select if the VCR will output on channel 3 or channel 4 (depending on which is vacant in your area). This switch and the TV must be set to the same channel.

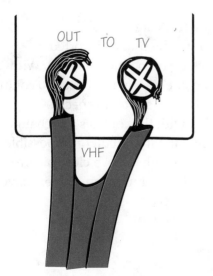

Step 5-18.
Checking the cables and connectors.
If there is no picture or no sound, yet the television is definitely on and functioning, check the cables and connectors. If they are connected correctly, you might have to remove the cable and test them. (Chapter 9 will show you how.)

Step 5-19.
Checking for programming errors.
If a timed or programmed recording didn't work, chances are nearly 100% that you did something wrong when setting the controls. Look over the various programming controls and study the owner's manual of your particular machine.

Step 5-20. Adjusting the tracking.
If the tape is playing but the picture is distorted or streaked, adjust the tracking (in both directions). With some of the newer VCRs, the information used to control tracking and other functions is stored in special electronic chips. Fiddling with the controls can change the "default" settings, making it difficult or impossible to adjust the tracking far enough to get a clear picture. The cure in such a case is to unplug the machine. Leave it unplugged for at least 30 seconds.

Step 5-21. Abandoning an overpowering copyguard signal.

Also for a distorted picture, eject the cassette and examine it. The encoded copyguard might be so strong that your VCR will play the tape, but the VCR and or TV will receive an over-powering copyguard signal that ruins the playback. Try another cassette and see if it cures the problem.

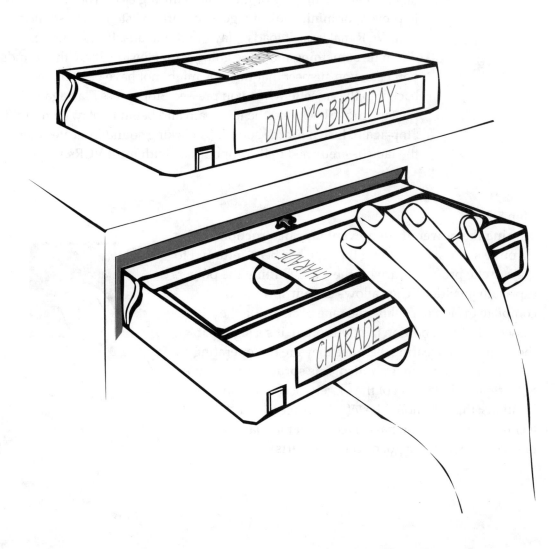

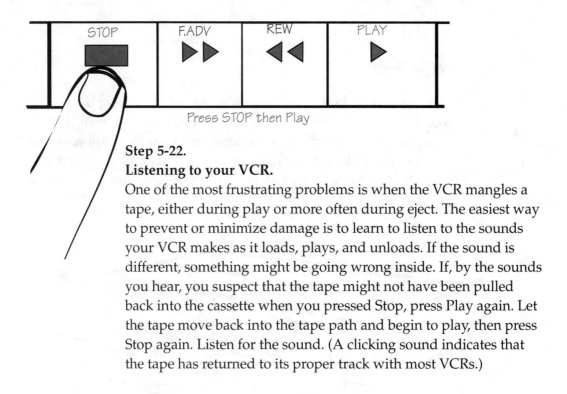

Press STOP then Play

Step 5-22.
Listening to your VCR.
One of the most frustrating problems is when the VCR mangles a tape, either during play or more often during eject. The easiest way to prevent or minimize damage is to learn to listen to the sounds your VCR makes as it loads, plays, and unloads. If the sound is different, something might be going wrong inside. If, by the sounds you hear, you suspect that the tape might not have been pulled back into the cassette when you pressed Stop, press Play again. Let the tape move back into the tape path and begin to play, then press Stop again. Listen for the sound. (A clicking sound indicates that the tape has returned to its proper track with most VCRs.)

Step 5-23.
Turning the power off to prevent tape destruction.
If you're still not certain that the tape has been returned to the cassette properly, turn off the VCR. Cutting the power, especially in older VCRs, allows certain springs to complete the job of pulling the tape back inside the cassette. If this condition happens often, it's a sign that the springs are wearing, because they are not operating reliably while tension is applied. The springs might have to be replaced. Because of the variety of springs and their locations, this job should be left to a professional technician (unless you have a technical manual for your particular VCR and a good source of parts).

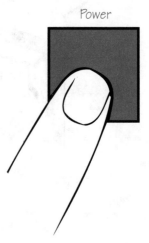

Power

Step 5-24. Watching as the VCR plays.

If the problem of tape eating is persistent, open the VCR and watch it go through a load, play, and unload sequence. Do this several times, watching for things that might be causing the problem.

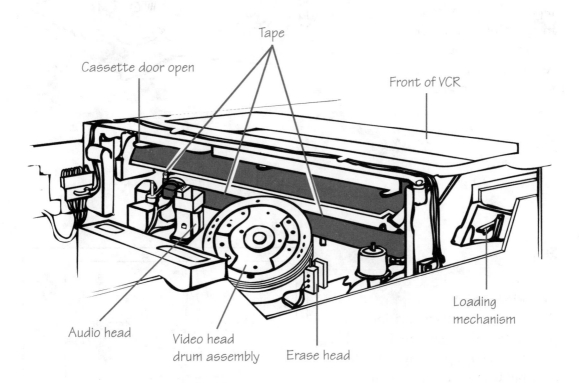

Tape

Cassette door open

Front of VCR

Loading mechanism

Audio head

Video head drum assembly

Erase head

Step 5-25.
Checking the idler assembly.
Many VCRs use an idler assembly
with a rubber tire on a spring-
loaded wheel (see chapter 7).
If this tire is worn or if it is
becoming overly dry, the VCR
might put most, but not all, of the
tape back into the cassette. Look
inside your VCR to see if it has
something that looks even
remotely like the illustration. If
so, turn to chapter 7.

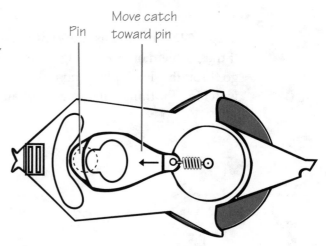

Pin   Move catch
        toward pin

Step 5-26.
Checking the TV.
Don't forget that the television is a
part of the overall system. It could be
that the VCR is operating perfectly
and that the problem is with the
television set. Is it on? Does it work?

Step 5-27. Changing cassette lengths.

The standard VHS cassette can hold up to six hours of video. People demanded more, which brought about the T-160 cassette. This cassette can hold up to eight hours of video. It uses a tape that is more thin than usual. This thin tape might be moving through the path less reliably. If you are using these tapes, try switching to the standard cassettes.

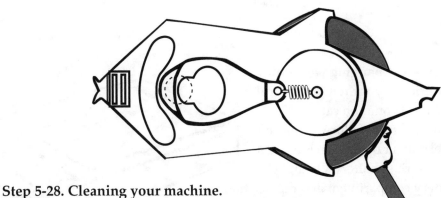

Step 5-28. Cleaning your machine.

Another common cause of VCR malfunction is excessive filth inside the machine. You might or might not be able to see what is causing the problem. Either way, thoroughly clean the inside of your VCR, including the entire tape path, and the video and audio heads (see chapter 5).

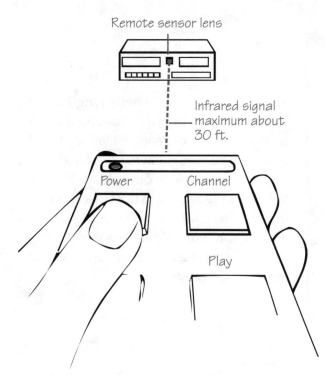

Remote sensor lens

Infrared signal
maximum about
30 ft.

Power   Channel

Play

Step 5-29.
Checking your remote control unit.
These days almost every VCR comes with a wireless remote control. Some use ultrasonics, but most operate using infrared signals. Both are invisible and perfectly safe. Infrared-type remote controls operate in line-of-sight. The signal can be blocked if something is between the remote control and the VCR. Excessive distance between the two might also prevent operation.

Step 5-30. Changing worn-out batteries in your remote control.
The most common cause of remote control failure, including intermittent operation, is a worn (or wearing) battery. Open the battery compartment and replace the battery (or batteries). Pay close attention to the polarity (the + and – labels). With single-cell batteries (AAA, AA, C and D, etc.—although the latter two are never used in a VCR remote control) the positive (+) side is the post; the negative (–) is the flat side.

Step 5-31. Cleaning the lenses.

In rare cases, the lens of the remote control unit, or the receiving lens on the VCR, might be so dirty that the signals are partially blocked. Clean the lenses with a clean, soft cloth and a cleaning fluid safe for plastic.

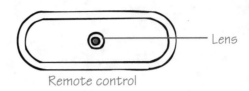

Remote control

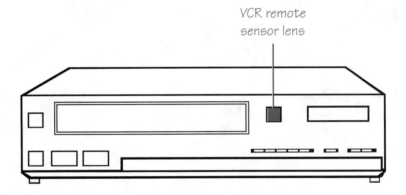

Shopping List for All Thumbs Guide to VCRs

- ❏ Volt-ohmmeter
- ❏ Needle-nose pliers
- ❏ Medium Phillips screwdriver
- ❏ Standard flat-blade screwdriver 3/8″
- ❏ Isopropyl alcohol (at least 95 percent purity)
- ❏ VCR cleaning solvent (optional)
- ❏ Swabs or optical-grade chamois
- ❏ Spray cleaner (optional)
- ❏ Splicing block
- ❏ Splicing tape
- ❏ Scissors
- ❏ Regular pliers
- ❏ Cartridge
- ❏ Small paper cups
- ❏ Adhesive tape
- ❏ Lint-free gloves
- ❏ Dubbing (copying) cables
- ❏ _____
- ❏ _____
- ❏ _____
- ❏ _____
- ❏ _____

Needle-nose pliers

Medium Phillips screwdriver

Standard flat-blade screwdriver 3/8″

Volt-ohmmeter

ISOPROPYL ALCOHOL 99% PURE

SPRAY CLEANER

CLEANING SOLVENT

SWABS

Splicing block

Regular pliers

Splicing tape

Lint-free gloves

Dubbing cables

Cartridge

Before you start

o Think safety.

o Work slowly and carefully.

o Read and understand all instructions.

o Gather all your tools and required materials.

o Make sure your test equipment is working properly.

Limitations

The most important factor is safety—yours first and then the safety of the VCR. Nothing is more important!

A close second is to realize your limitations. The purpose of this book is to help you understand that you have fewer limitations than you suspected. Now it's time to turn that around.

What if cleaning the machine doesn't cure the problem? You can try a few more things, as detailed in the book. Many times you'll be able to fix the problem, or at least spot the cause. Approach the problem with confidence, and realize that in most cases you can do it.

However, only you can tell when you've reached your limit. Be honest enough with yourself both so that you don't stop too soon, and so you don't stop too late.

Eventually the point is reached when you face the choice of bringing the VCR to a professional, or throwing it in the trash and buying a new one.

Warranties

In many cases if you open the cabinet while the VCR is under warranty, you could void that warranty. Fortunately, not much is likely to go wrong during the warranty period.

If the VCR is still under warranty, it's generally best to let the authorized service center take care of any problems. The warranty generally covers everything except cleaning.

Obviously this means that you will have to know where authorized service centers are. Often you'll get a list of authorized centers, and sometimes phone numbers to call when you buy your VCR.

From *All Thumbs Guide to VCRs* by Gene B. Williams
© 1992 by TAB Books, a division of McGraw-Hill, Inc.

CHAPTER SIX

Cleaning Your Machine

Mechanical things have a common enemy—dirt. If a machine is allowed to get dirty and stay dirty, it won't operate correctly. It could stop working, and it might suffer permanent damage. When VCRs are kept clean, they work better, last longer, and have fewer problems.

During recording or playback, the video and audio heads of your VCR are in direct physical contact with video tape. The manufacturing processes for tape have improved, but there is no such thing as perfection. The physical movement of the tape across the parts leaves behind tiny particles. These particles build up over time. Unless you clean them away, the deposits can cause real trouble.

The key is preventive maintenance. Protect your VCR the best you can. Give it a thorough cleaning on a regular basis.

Tools & Materials

- ❏ Isopropyl alcohol
- ❏ Liquid Freon
- ❏ Swabs
- ❏ Cloth
- ❏ Needle-nose pliers
- ❏ Screwdriver

The alcohol must be "technical grade," which means that it must be of at least 95% purity, and preferably higher. The cleaning solvent can be used for cleaning all parts, but it is essential for cleaning the rubber parts. The swabs must be of the proper materials and have no threads.

Step 6-1. Cleaning the cabinet and loading platform.
Clean the cabinet with a cloth to reduce the amount of dust that seeps inside and to keep the VCR looking more attractive. Also clean the loading platform with the cloth; use a swab to clean the corners. Your goal is to be sure that the inserted cassettes do not drag contaminants into the machine. Keep the VCR covered when not in use, but remove the cover when you are using the machine.

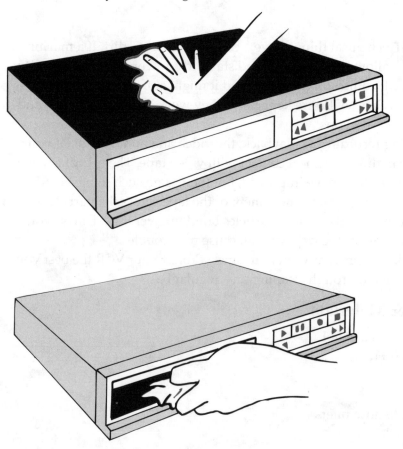

Step 6-2. Getting inside.

The instruction manuals that come with VCRs do not tell you how to get inside them. (*Caution*: Opening the cabinet while the VCR is still under warranty could void that warranty!) With most VCRs, the top cover is held in place by two or four screws. If only two screws are used, they will be at the back. These screws generally go down through the top, but there also might be screws in the back or sides. Remove the screws carefully and put them in a safe place. Lift at the rear while gently pulling backwards to release the lip that holds the cover at the front of the VCR. If the cover doesn't lift easily, carefully inspect the VCR for screws you've missed. (In a few cases, there might not be any holding screws on top. The screws might be in the back or more rarely through the sides.)

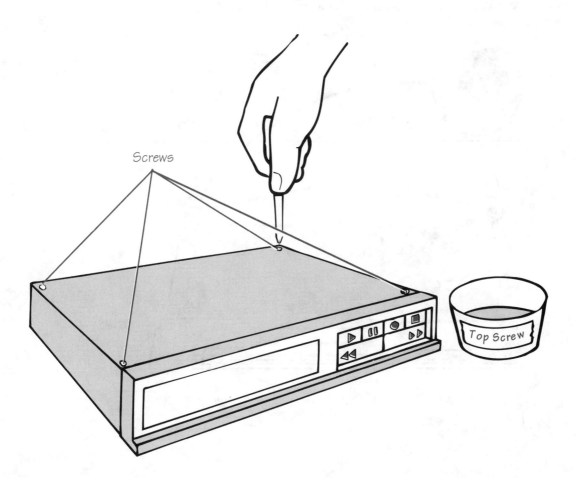

Screws

Top Screw

Step 6-3. Removing the shield.

Inside some VCRs have a metal shield over the head assembly and tape path, which must be removed to get at the parts beneath. The shield is usually held in place by two screws and two thumb catches. Other VCRs have a partial shield or lack a shield entirely. Remove the screws and put them in a safe place. Be careful not to let them fall into the VCR. (The screws are very small so it's easy to lose your hold on them. You can use a pair of needle-nose pliers to hold the screws while removing them. Do not squeeze hard because you can damage the threads.) Use your thumbs to release the two thumb catches. Do not force them. If the catches do not release, try pressing gently downward on the shield while working the catches. Carefully lift the shield; there are usually delicate wires near by. Once you've removed the top cover and shield (if any), you can get at the audio head, video head, and tape path.

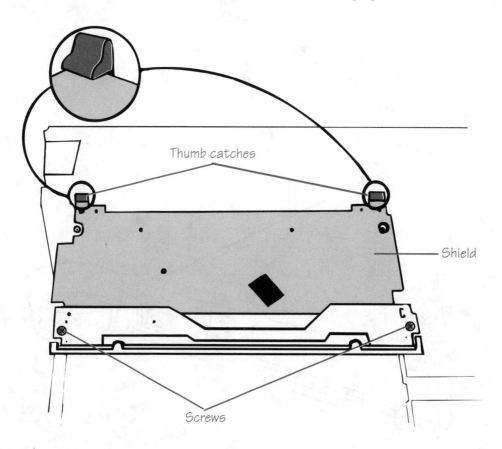

Thumb catches

Shield

Screws

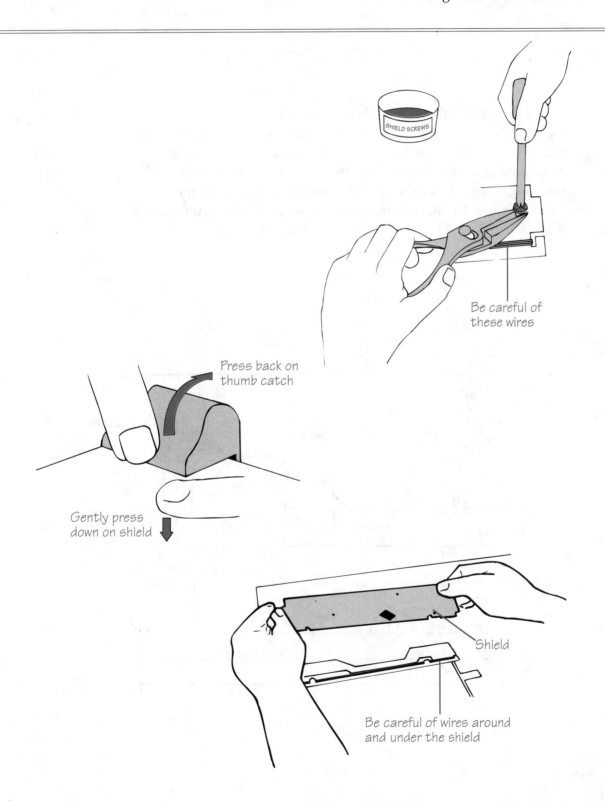

SHIELD SCREWS

Be careful of
these wires

Press back on
thumb catch

Gently press
down on shield

Shield

Be careful of wires around
and under the shield

Step 6-4. Cleaning the video head assembly.

Wet a swab with cleaning fluid and then press it against the inside of the bottle to squeeze out the excess. (Do not use cotton swabs on the video head assembly.) Gently hold the swab steady against the assembly while rotating the assembly with your finger. Do not apply much pressure. Let the cleaning fluid do the work, not the pressure. It is best if you wear a lint-free glove when doing this work.

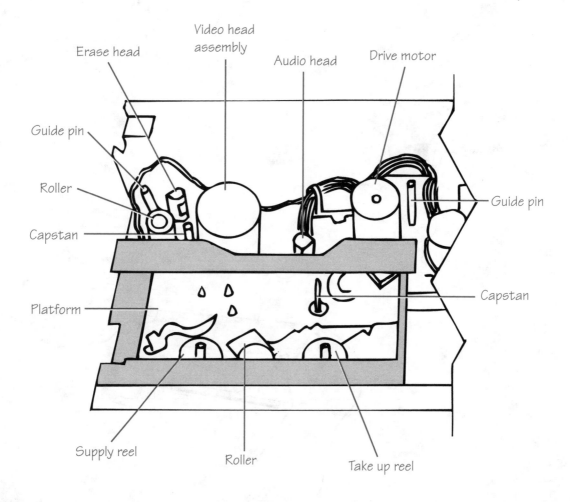

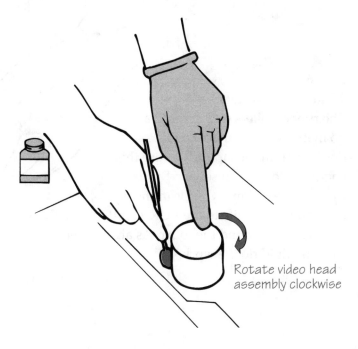

Rotate video head
assembly clockwise

Step 6-5.
Cleaning the audio and erase heads.
If the audio and erase heads are separate
from the video head assembly, they
should be cleaned too, again, the key is to
be gentle and to let the cleaning fluid do
the work. Do not scrub! (*Caution*: Do not
use alcohol on rubber parts! Alcohol will
cause them to dry out and decompose.)

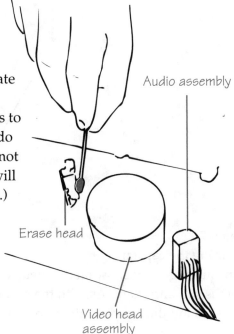

Audio assembly

Erase head

Video head
assembly

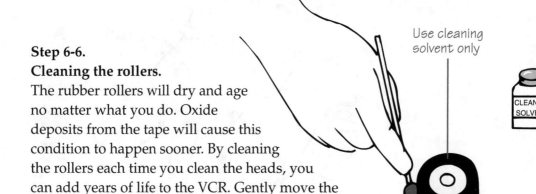

Use cleaning
solvent only

CLEANING
SOLVENT

Rubber roller

Step 6-6.
Cleaning the rollers.
The rubber rollers will dry and age no matter what you do. Oxide deposits from the tape will cause this condition to happen sooner. By cleaning the rollers each time you clean the heads, you can add years of life to the VCR. Gently move the wetted swab around the roller. As always, let the cleaning fluid do the work.

Clean around and
up and down

Step 6-7.
Cleaning the capstans.
Deposits might not show on the metal capstans, but cleaning them is still important. As with the rollers, move the wetted swab around the capstans, then up and down.

Capstan

Step 6-8.
Cleaning the platform.
While the VCR is open, slightly dampen a
swab with isopropyl alcohol or cleaning
fluid and clean away dust and lint from
the platform and other parts. Be careful to
remove just this and not the factory
lubricants.

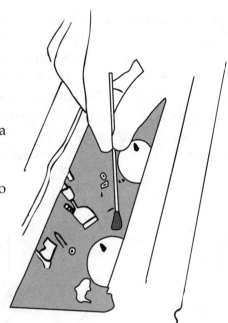

Step 6-9. Cleaning the cable connectors.
Sometimes the only problem is that the cable connectors need to be cleaned.
Larger connectors, such as those that connect the VCR to the television set,
can be cleaned with alcohol or cleaning fluid and a swab. Some
connectors, such as those for circuit boards, have delicate
pins. You should use a spray cleaner for these
so that the pins don't get bent or
damaged. It's relatively rare that
the connectors need cleaning. Do
so only when necessary.

Cable connector

Step 6-10. Cleaning circuit board connectors.

To remove a circuit board connector, first inspect it for any catches, holding screws, etc. Grasp it with a finger and thumb. Hold the circuit board with your other hand to prevent it from moving. Pull gently; do not force it! To spray clean the connector, spray away from yourself. Use the spray sparingly. Don't forget to clean the connector on the circuit board. (*Caution*: Do not use a spray cleaner with any kind of lubricant in it. Read the label carefully.)

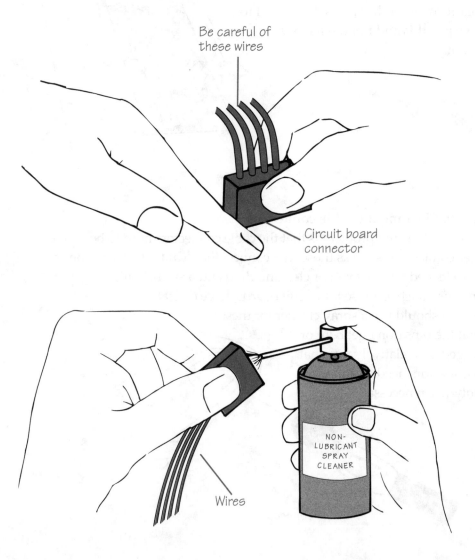

Step 6-11. Cleaning a connecting strip.
A few circuit boards, usually in older VCRs, might have a connecting strip rather than pins. Cleaning these is much the same as cleaning pin-type connectors. You can use a swab and fluid or a spray cleaner.

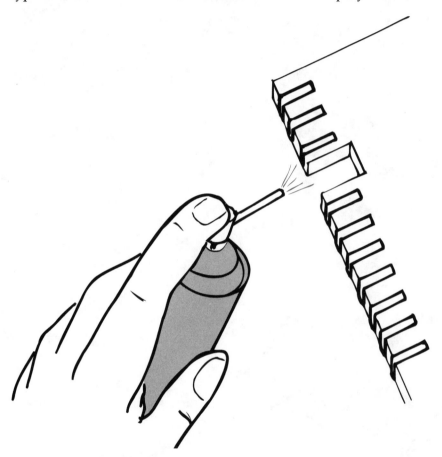

CHAPTER SEVEN

Idlers, Belts & Other Parts

T hings work in a certain way for a specific reason. When something fails, there is a reason for it. For example, if a motor drives a device and the two are connected via a belt between two pulleys—and if this belt breaks—it doesn't matter if both the motor and device are in perfect condition, because neither can perform its function without that belt.

The placement of a particular part might be different in your VCR, but its function is often very much the same. In many cases the diagnosis is little more than a matter of looking. A broken belt can be easily spotted. Even if you've never seen the inside of the machine, figuring out where the broken belt came from is usually just a matter of inspection.

The same is true of other problems. A part might be obviously broken or bent. A roller might be discolored from excessive oxide buildup or it might be cracking indicating age and wear. Even the components and circuit boards sometimes reveal the cause of their malfunction.

You should first determine the symptoms—what is happening or not happening. If you think it will help, write a list of the symptoms, along with your best guesses as to what might be causing the problem. This will help you to eliminate where the problem is not.

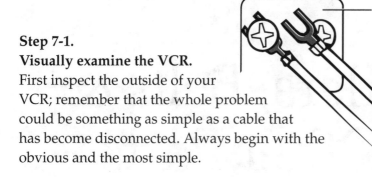

Step 7-1.
Visually examine the VCR.
First inspect the outside of your
VCR; remember that the whole problem
could be something as simple as a cable that
has become disconnected. Always begin with the
obvious and the most simple.

Step 7-2. Remove the cabinet and look inside.
Remove the cabinet as instructed in Step 6-2. Visually examine the
internal components of the unit. In this exaggerated example, notice
that a wire on one of the connectors is broken, that a capacitor on a
circuit board has broken open and burned, and that somehow a paper
clip got inside.

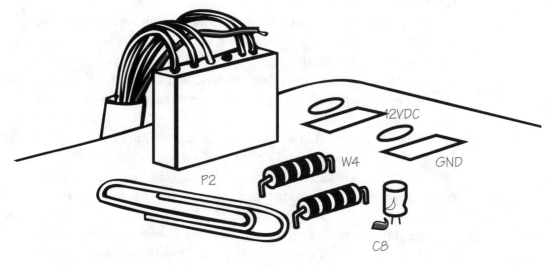

Many VCRs use an *idler assembly* to help control tape movement. This small device is usually mounted on a pin between the two reels. The idler assembly loosely flips back and forth between the reels. If it wears, tape movement might not be reliable. The result is often mangled tapes. Most idlers have a rubber tire; some use a geared wheel instead. If yours has a gear, the following steps can be skipped.

Step 7-3.
Cleaning the idler assembly.
Using a swab and cleaning fluid that is safe for rubber, clean the rubber tire. Rotate the wheel and tire with a finger while gently scrubbing the tire with the wetted swab.

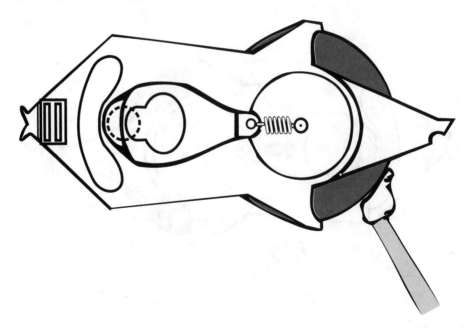

Step 7-4. Removing the idler assembly for changing or cleaning.
Remove the idler assembly with your fingers or a pair of needle-nose
pliers. The most common kind of mounting is to have a spring-loaded
clip that holds the device in place. To release the clip, push the catch
toward the pin. While holding this catch in, carefully lift the idler
assembly upwards. Cleaning the tire is easier and usually more
thorough with the assembly removed.

If cleaning doesn't solve the problem, replace the tire. An exact match
is needed. Parts supply stores that carry idler tires can usually make this
match by either the make and model of the VCR or by physical
measurement. The make and model can be found in your owner's
manual or on a sticker located either behind the VCR or beneath it.

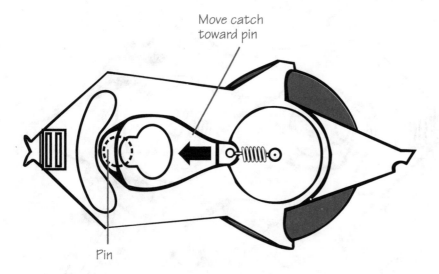

Move catch
toward pin

Pin

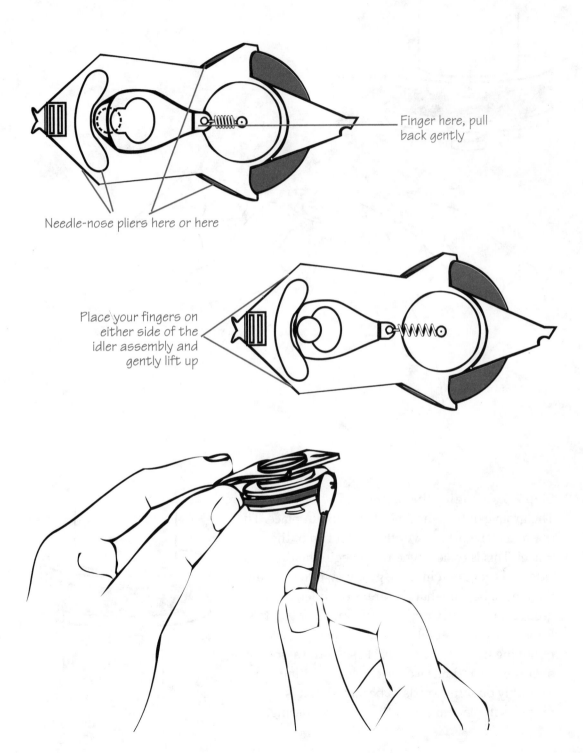

Finger here, pull
back gently

Needle-nose pliers here or here

Place your fingers on
either side of the
idler assembly and
gently lift up

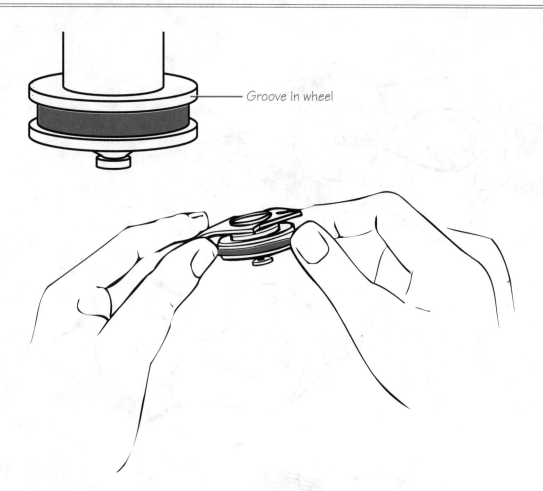

Groove in wheel

Step 7-5. Changing the idler tire.

The groove in the wheel holds the tire in place. The tire must be worked over the wider rim of the wheel. This is easiest when you begin the tire on one side and squeeze it into the groove while rotating the wheel. Be sure that the tire is completely in the groove and is flat. Check for any kinks or twists. Replacing the assembly is just the reverse of removing it. Push the clip back, slide the assembly over the pin and release the catch. Try lifting the assembly out afterwards to be sure that it is properly in place and that the catch is holding.

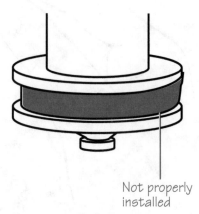

Not properly installed

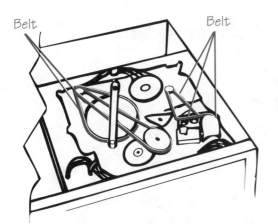

Belt   Belt

Step 7-6.
Locating the belts in your VCR.

Most of the belts and motors are located beneath the VCR. Wherever they are located their job is to transfer the action of a motor to some device. The belt fits into the groove of a pulley on the motor, and into the groove of a pulley on the device. To get to the belts beneath your VCR, you have to remove the lower cover. To do this, first very carefully turn over the VCR.

Step 7-7. Removing the bottom cover.

The bottom cover is held with 4, 5, or 6 screws. The most common pattern is to have a screw in each of the four corners and one near the center. Remove the screws and put them in a safe place. If you also still have the top cover and shield off be sure that each set of screws is labeled. Once the screws are removed, the bottom cover will lift off easily. If it doesn't, examine the VCR thoroughly to find out what you've missed. Don't force anything. With the bottom cover removed, you can get at the belts, motors, etc.

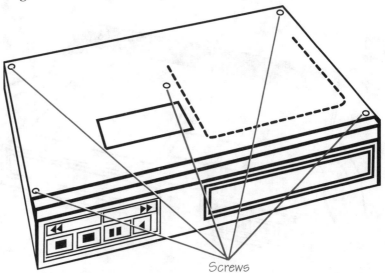

Screws

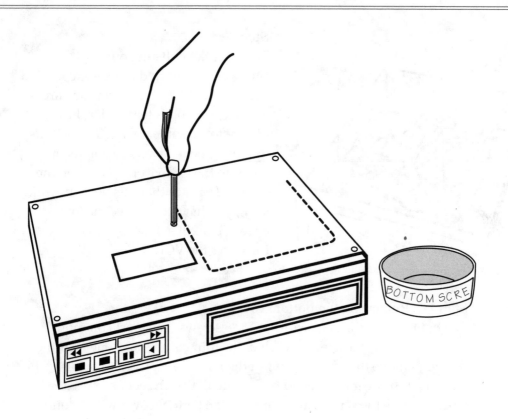

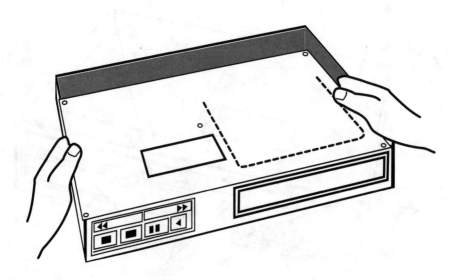

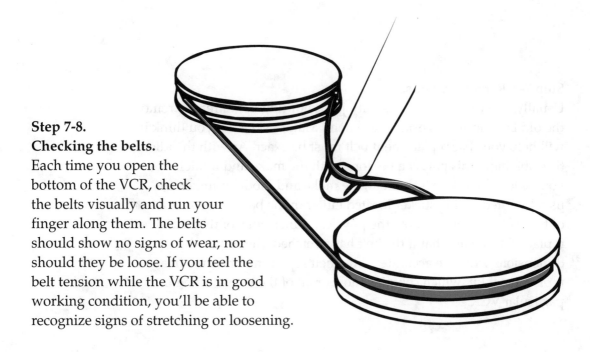

Step 7-8.
Checking the belts.
Each time you open the
bottom of the VCR, check
the belts visually and run your
finger along them. The belts
should show no signs of wear, nor
should they be loose. If you feel the
belt tension while the VCR is in good
working condition, you'll be able to
recognize signs of stretching or loosening.

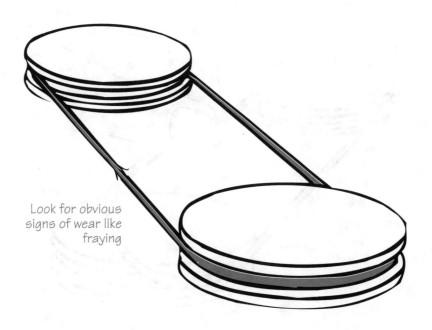

Look for obvious
signs of wear like
fraying

Step 7-9. Removing belts.

Usually, you can change belts with just your fingers. Carefully remove the old belt, noting its exact placement. Make a sketch if you think it will help you. The replacement belt must be exact. As with the idler tire, obtaining this part can be done with the make and model. If you've lost the owner's manual, the make and model number are usually listed on the sticker located either in the back or on the bottom of the VCR, or you can note the physical dimensions of the belt to be replaced. Be aware that if the belt has stretched, the physical dimensions won't be accurate. Often belt replacements come in a replacement kit, which contains most or all of the belts needed in a particular VCR.

Step 7-10. Changing belts.

To install the replacement belt, you can usually just stretch it into place. If you have difficulty, set the belt in the groove of one pulley, start it in the groove of the other, and rotate that second pulley so that the belt slides into the groove. Be sure that the belt is securely in the grooves of both pulleys and that there are no twists in the belt.

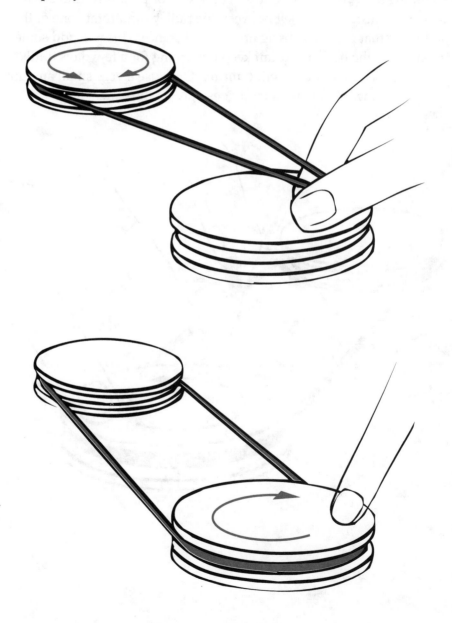

If you do not have access to a parts store, you can temporarily cure a slipping belt with a piece of resin, beeswax, or paraffin. Hold the wax against the inside of the belt between the pulleys and rotate one of the pulleys so that the belt rubs against the substance. This will add some stickiness to the belt so you can keep it working for a few days until you can get a replacement belt. *Caution!* Only use a very small amount of the substance, and make sure it goes only on the belt.

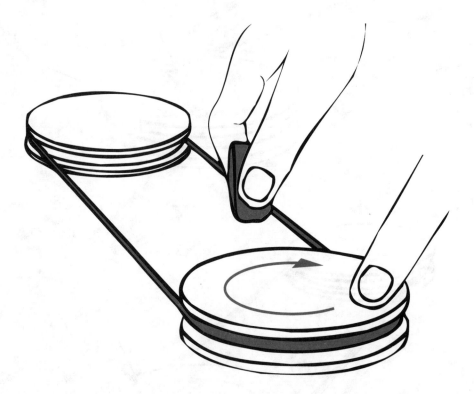

CHAPTER EIGHT

Tapes

Cassettes are an integral and important part of the video system. They are also one of the most delicate parts of the video system and the one most likely to cause trouble. Many things can go wrong with a cassette or the tape contained in it. Cassettes can be manufactured poorly, which can mean anything from a warped case or low-quality tape to a missing spring that was never installed. Even the best-made cassettes wear in time. Accidents like spilling liquid on a cassette or exposing it to heat also can cause problems.

Bad tape can cause damage to the VCR by leaving excessive deposits or having a "sandpaper" effect on your machine. A tangled tape also can cause damage, especially if you become impatient and begin yanking on it. Rental and borrowed tapes bring yet another potential danger to your VCR. As a tape goes through a VCR it leaves behind small deposits, and it also can pick up dirt, dust, and grime. These can be transferred into the next VCR that uses the tape. In other words, whatever is in someone else's VCR can be moved into your VCR.

Protecting your investment begins by buying only quality cassettes. You don't have to buy the most expensive cassettes available. Tests have shown that at times quality has little to do with cost. Buying quality tapes means staying with the known brands and avoiding the off-brands or store brands. Words like "supreme quality" or "professional" don't necessarily mean anything. In general, if you see a tape being offered by only one store, it's not what you're looking for. A brand that can be found almost anywhere is more likely to be reliable.

The best cassette won't last long if you abuse it. Cassettes are sensitive to magnetism, heat, cold, moisture, dust and other things. Your task is to protect the cassette as best you can. Keeping your VCR clean helps you accomplish this task (see chapter 6).

Step 8-1. Rewinding the tape.

Completely rewind the tape before storage. Also some experts suggest that if a tape has been stored for a long period of time, completely fast-forward and then rewind before playing it to repack the tape inside the cassette. Although this much effort isn't always needed, it is sometimes a good idea.

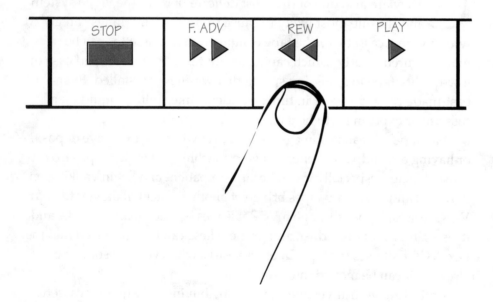

Step 8-2. Keeping your cassettes in cases.
The door side of the cassette should be inserted into the case first. The
most vulnerable part of the cassette is then deepest into the box where
it is best protected. Special hard boxes are available that seal the
cassette inside more completely. To further protect your tapes, store
them in a VCR case or cabinet. Open storage racks are fine for storing,
but they do little to protect the cassettes.

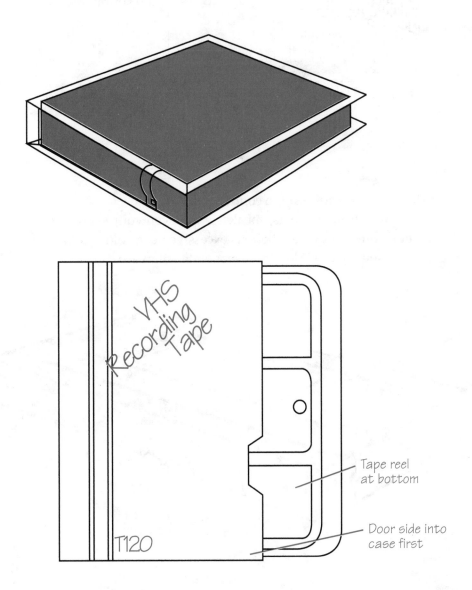

VHS
Recording
Tape

T120

Tape reel
at bottom

Door side into
case first

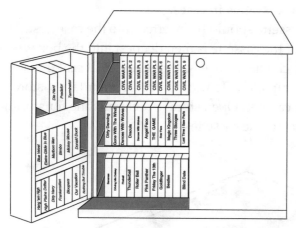

Step 8-3.
Storing cassettes on end.
Store your cassettes vertically with the supply reel at the bottom. Do not stack them flat.

Step 8-4. Avoiding troublesome conditions.
Keep tapes away from magnets, speakers, and even your telephone. Also, do not store tapes where there is excessive heat, cold, sunlight, or moisture. Cassettes exposed to these elements will warp.

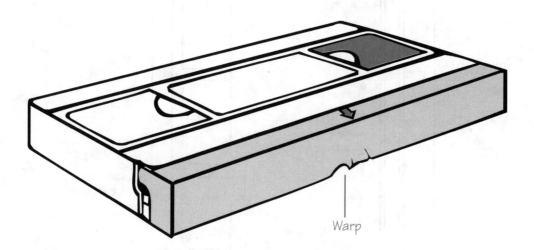

Warp

Although there are slight differences in manufacture and construction, most cassettes are very much the same. The top of the cassette has a space for the label and clear viewing windows. The square button releases the catch for the door.

Door release button

The bottom of the cassette has various notches and holes. Note in particular the brake release (reel release) hole and the locations of the screws that hold the cassette together.

5 screws

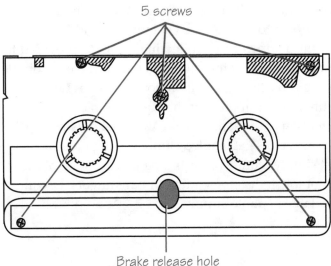

Brake release hole

The inside consists of a supply reel, a take-up reel, guide rollers, springs, and a locking "paddle" that controls tape direction. The cassette also might have a thin sheet of plastic over the tape to give the tape a smoother, slicker surface on which to move.

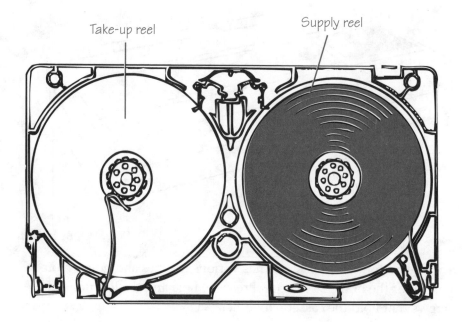

Take-up reel

Supply reel

No matter how well you care for your VCR and tapes, sooner or later you're going to have a tape get tangled inside. You can reduce the chances of this happening by following the steps provided for proper storage and care of cassettes, but you won't be able to completely eliminate the problem, especially if you rent tapes often.

Sometimes you can prevent the tangle from ever happening by paying attention to the sounds your VCR makes (see Step 5-22).

If your tape does get tangled, straightening it out usually isn't difficult; it simply requires care and patience. For all of the following steps, do not touch the tape with your bare fingers. Use lint-free cotton gloves available in photo supply stores.

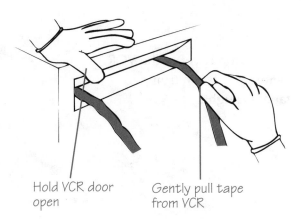

Hold VCR door open

Gently pull tape from VCR

Step 8-5. Extracting the loose tape.
If you eject the tape and see that the tape is still outside the cassette and it is now caught, do not pull. Open the door to the VCR and look inside. there might be loose tape that you can pull out easily.

If you can't tell if the tangle is serious or not, very gently pull the tape to see if it will come out of the VCR. If it doesn't come out easily, stop! Go to Step 8-7.

Step 8-6. Manually rewind the cassette.
If the tape comes out of the VCR with little damage done to it, you can simply wind it back inside the cassette. Press the door-release catch on the side of the cassette and prop it open with a pencil or similar object. Insert your finger into the reel at the bottom of the cassette and turn it to bring the tape inside the cassette. If one reel doesn't turn, try the other one. If neither reel moves, you might have to insert a pencil into the reel (brake) release hole.

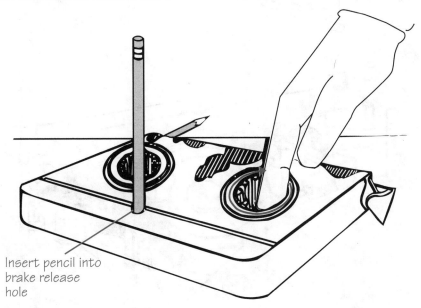

Insert pencil into brake release hole

Step 8-7. Opening the VCR.

If the tangle is bad, you will have to remove the top cover of the VCR. Visually examine the inside of the VCR. Before you begin trying to untangle the tape, try to spot where the tape is caught.

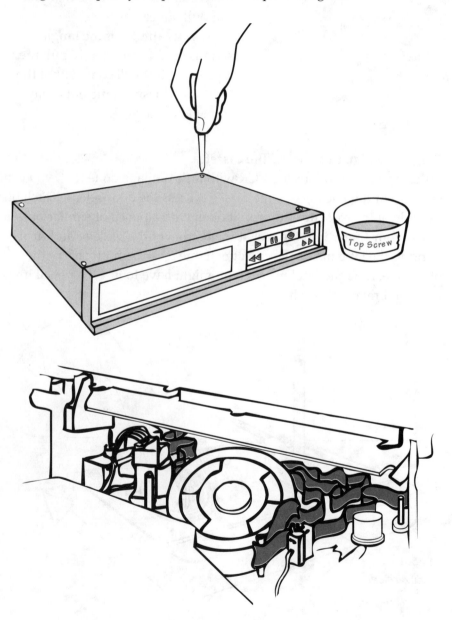

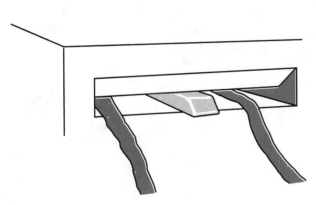

Step 8-8.
Propping open the doors
of the cassette and VCR.
If the tape doesn't come out
easily, you might need both
hands to work with the tape.
Letting the door close on the tape
can cause additional damage. An
eraser or similar object can be
used to hold the door open so
that both of your hands are free.

Step 8-9. Extracting the tape from the VCR.

Very gently get the tape loose and pull it out of the VCR. Then, as in
Step 8-6, wind it back inside the cassette. It's important that you do all
of this gently and keep the tape as straight as possible. Never force or
yank on the tape.

If there was little or no damage done to the tape, you can probably
get by with using the Fast Forward and Rewind controls to bring the
tape all the way to each end. (Doing this twice is advisable just to make
sure the tape is packed tightly on the reel.)

If the damage is more severe and the recording is so valuable that
you cannot scrap it, you have three ways to repair it: you can make an
in-tape repair, a repair at the leader, or a tape swap. Each method has
advantages and disadvantages. In all, tape repair is a last resort when
the recording is too valuable to lose because you risk causing damage
to the VCR. You should attempt repair only when the recording is of
more value to you than the VCR itself.

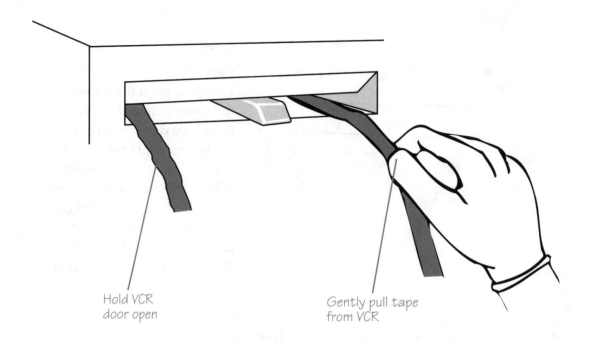

Hold VCR
door open

Gently pull tape
from VCR

If a repair is necessary, make a copy of the recording immediately. Set the original aside, to be used only in an emergency. You'll lose some quality in playback with the dub, but you'll protect your VCR. Each time the repaired sections move across the video head, you risk causing permanent damage to the VCR.

An in-tape splice is the simplest but least efficient method of video tape repair, because the place where the two pieces of tape are joined together passes across the video head. If you decide to use this method, follow Steps 8-10 to 8-12.

Making the repair at one of the two clear leaders is safer than making an in-tape repair, because it reduces the chances of the splice going across the video heads. You can save all of the tape before or after the damaged section with this method, but if you need the whole recording from beginning to end, this option would not work. See Step 8-14 for instructions on this method.

When you want to avoid an in-tape splice and still keep the original recording intact, you can purchase a second cassette for a tape swap. See Steps 8-14 to 8-17.

After you repair a tape, regardless of which method you use, you should make a copy of the original. See Step 8-18 for details.

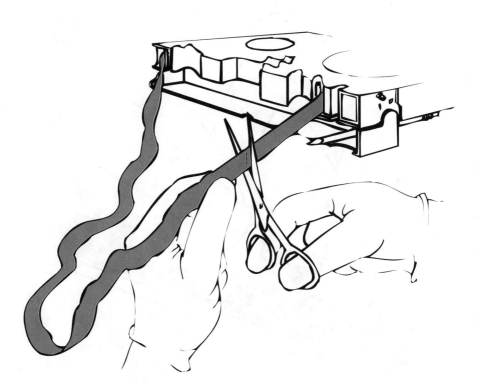

Step 8-10.
Removing damaged
tape for an in-tape splice.

Prop open the front door of the cassette with a pencil. If necessary, insert another pencil into the reel-release hole located on the bottom of the cassette. Pull some tape from the cassette. Carefully cut away the damaged section. It's best to use a splicing block if you can find one that is suitable for 1/2-inch video tape. Overlap the two ends of video tape and carefully slice through both with a sharp razor blade. Your goal is to slice the two loose ends so that they match exactly. Many professionals prefer a 45 degree angle because it provides more surface area for the splicing tape to cover. Brush away the trimmed excess.

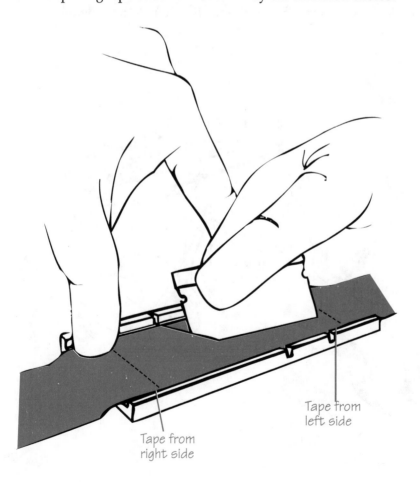

Tape from
right side

Tape from
left side

Step 8-11.

Adhering the ends together.

Apply splicing tape to the nonoxide side (inside) of the video tape. Be sure that both ends of the video tape are the same width, that the inside of both is up, and that the tape is not twisted. Actual 1/2-inch splicing tape is preferred because it matches the width of the video tape exactly; however, it is hard to find. A quality adhesive tape (cellophane, such as Scotch) can be used if the splicing tape cannot be found. The ends of the video tape must match exactly so that there is no splicing tape in a gap between the two ends or beyond the edges. If necessary, carefully trim the excess splicing tape; using a splicing block makes precise trimming much easier to accomplish.

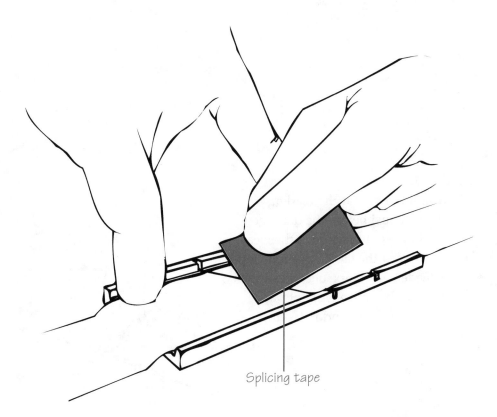

Splicing tape

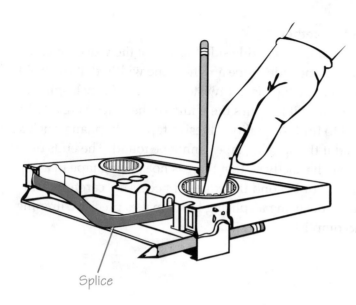

Splice

Step 8-12.
Winding the tape
back in the cassette.
Carefully wind the spliced tape inside the cassette. Again fast forward and rewind to each end of the tape at least once and preferably twice to firmly pack the tape on the take-up reel.

Step 8-13.
Making the repair at the leader.
Cut the tape just before the damaged area and then cut the tape at the clear leader; discard the tape. Splice the tape to the leader. Your tape becomes much shorter, but the splice will not pass over the video head and risk damage to your VCR.

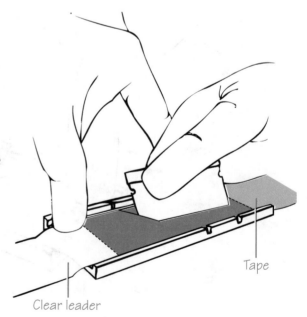

Clear leader

Tape

Step 8-14. Disassembling cassettes for a tape swap.

For this repair, you must obtain a new cassette. Disassemble both the new cassette and the damaged cassette by removing the screws from the bottom of the cassettes with a fine-tipped Phillips screwdriver, such as a jeweler's screwdriver. Make sure both halves stay together, because there are tiny parts inside that can fall out. Carefully turn both cassettes over with the label-side down. Separate the top and bottom halves.

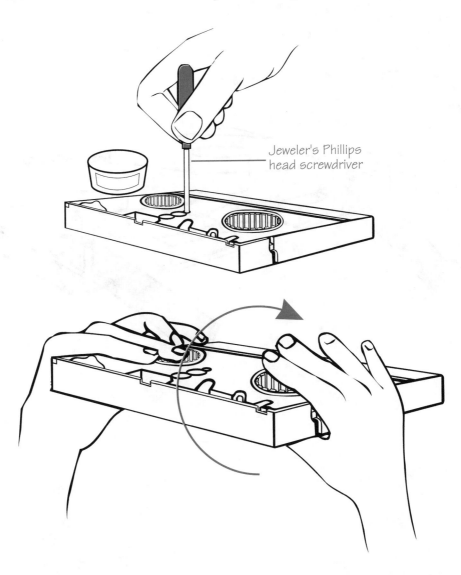

Jeweler's Phillips
head screwdriver

Step 8-15. Discarding the new tape.

Cut out and discard all of the tape in the new cassette at the leaders at each reel. Your goal is to get two empty reels with clear leaders.

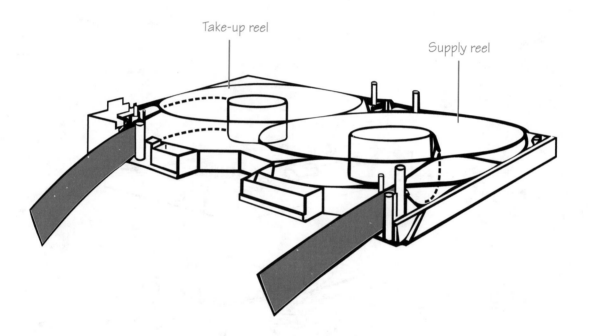

Take-up reel

Supply reel

Step 8-16. Swapping the reels.

As in Step 8-10, carefully cut away the damaged tape section. With both cassettes open, lift out the partially filled take-up reel and swap it with the empty reel of the new cassette. In each cassette, splice the tape to the clear leader as in Step 8-13. You now have the first part of the recording in one cassette and the second part of the recording in another cassette.

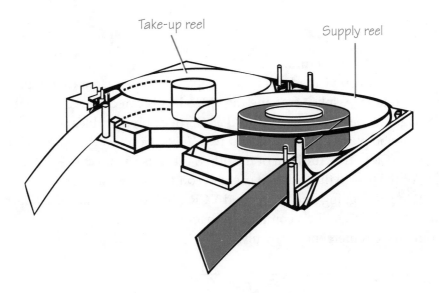

Take-up reel   Supply reel

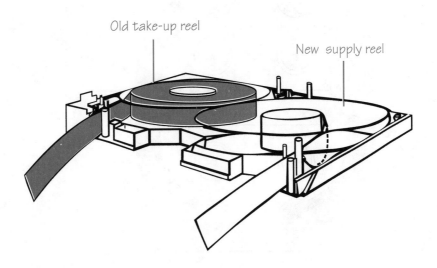

Old take-up reel   New supply reel

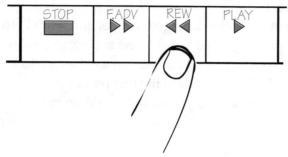

Step 8-17. Reassembling the cassettes.
Reassemble both cassettes by putting the top and bottom halves together again and replacing the screws. Fast forward and rewind both cassettes, again at least once and preferably twice.

Step 8-18. Making a dub.

Especially if you've made an in-tape repair, it is important to dub the repaired tape. A splice crossing the video heads can cause considerable damage. To make a copy, you need a pair of dubbing cables with RCA-type phono plugs on each end and two VCRs. Connect the Audio Out of one VCR to the Audio In of the other. Likewise connect the Video Out of the first to the Video In of the second. The signals from the first VCR are now going into the second VCR, allowing it to record what the first VCR plays. Make a complete dub and then store the repaired originals for emergency use only.

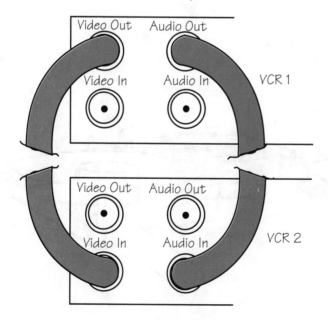

CHAPTER NINE

Connections

Alone a VCR is useless. To record, a VCR must be connected to an incoming signal; to play back, it must be connected to a television set; and if it is capable of stereo sound, the left and right channel outputs must be properly connected either to a stereo television or a home stereo system.

All of the basic cables are included when you buy a new VCR. These cables are adequate for a standard hookup. You can buy additional wires, cables, and connectors to make your VCR more versatile. For example, you might want to use two or more input signals (local antenna, cable, satellite dish, etc.). You might want a single VCR to drive two or more television sets (and at each of these, you might again want a selection as to which input the TV will show). You also might want to connect two VCRs together for dubbing taped recordings.

There are many different kinds of wire, cable, and connectors. What you need depends on your particular system and how you wish to use it. You can find different connectors and transformers at your local electronics store.

Step 9-1. Examining your VCR's input/output terminals.

Look behind your VCR for the input/output terminals. Most VCRs have three different kinds of terminals: the screwhead type, the threaded type, and a nonthreaded push-on type.

The screwhead type of terminal accepts the older 300-ohm twin lead from a television antenna.

The threaded type of terminal with screw threads and a small hole is for 75-ohm coaxial cable used with cable or satellite television and with a standard television antenna that has been equipped with a transformer (see Step 9-4).

The side-by-side push-on connectors with the larger hole have separate audio and video signals and are used when dubbing from one VCR to another, or when you wish to connect the VCR's audio (sound) to your home stereo system.

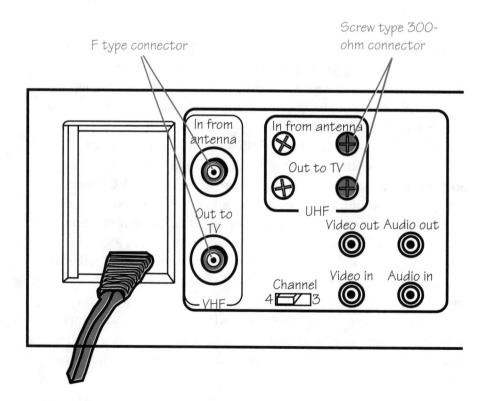

Step 9-2.
Determining your connectors.
Just as there are different inputs/outputs, there are different connectors on the wires or cables. The 300-ohm twin lead from a television antenna will have (or should have) u-shaped spade lugs that fit beneath the screws. The 75-ohm coaxial cable will have an F-typed connector (also called an RF connector). The connector used for dubbing or sending audio to the stereo is an RCA-type plug.

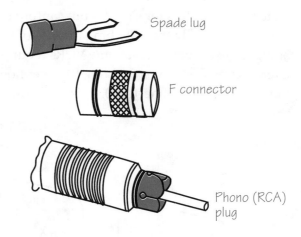

Spade lug

F connector

Phono (RCA) plug

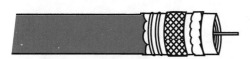

The F-type connector (also called an RF connector) is far more efficient. Like the 300-ohm twin-lead, this connector carries the entire (composite) video and audio signals, but through a coaxial cable that provides an impedance of 75 ohms. Once again, you don't have to worry about what this means, only that you cannot directly connect the 75-ohm cable to the 300-ohm antenna.

Note the connectors for signal input/output on the VCR. In most cases you will need only the 300-ohm or the 75-ohm in each direction—not both. The exception is that Some VCRs require a separate UHF input. Even if yours has inputs for both, try reception first with a single input through the VHF connector(s). This will probably work fine.

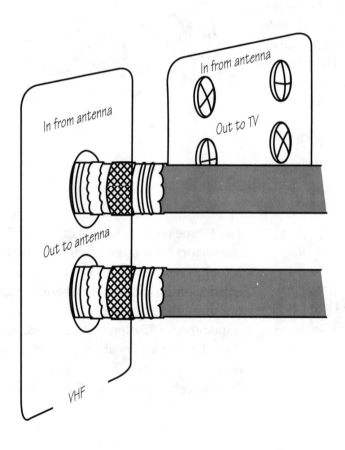

Step 9-3. Understanding cables and connectors.
Finally, not only must the connectors match, so must the wire or cable that carries the signal from the source to the input.

Spade lugs are used on 300-ohm twin lead. This wire consists of two stranded conductors that run side by side and are separated by insulating plastic. It has been used for a long time because a standard television antenna has a natural impedance of 300 ohms. You don't have to know what this means, only that you cannot directly connect a 300-ohm source to a 75-ohm input or vice versa. It is becoming less popular because it is prone to picking up interference, because it can't have any sharp bends along its route and because it has a shorter useful lifespan.

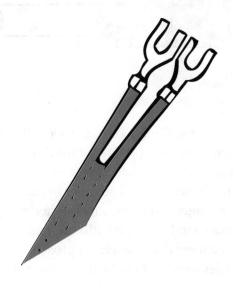

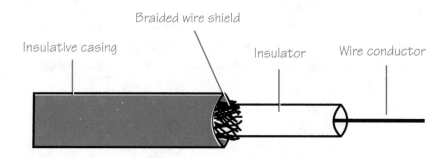

Far more efficient is the 75-ohm coaxial cable called R-59, R59U, RG59 or RG-59U. (You don't have to worry about the differences—they're all similar.) This has a solid center conducting wire, a surrounding insulator, a braided grounding shield wire and an outer insulation. Because the shield surrounds the center conductor, any interfering signal is sent to ground and can't reach (and interfere with) the wanted signal. Quality R-59 cable will have a more dense shield, approaching 100%. It is used for cable and satellite television, and can be used with a standard television antenna with a transformer (see Step 9-4).

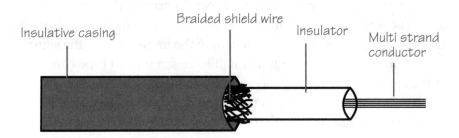

Insulative casing   Braided shield wire   Insulator   Multi strand conductor

It should be noted that you will be using either the 75-ohm input or the 300-ohm input (or output), but not both. The only time this is different is when the incoming UHF is split and separated from the VHF, and the only time this occurs is with some "rabbit ears" antennas. In most cases, even then you'll get both VHF and UHF on the VHF wires.

The cable used to carry the separated audio and video signals is similar to R-59 except that the center conductor is stranded rather than solid. Once again the center conductor is surrounded by an insulator, then the braided ground shield and an outer insulator. This cable is sometimes called "audio cable," and is usually smaller than R-59 cable. It's important to know that although the cables appear to be similar, they are not interchangable.

Step 9-4. Understanding RF transformers.

When you want to get the advantages of using coaxial cable to connect a 300-ohm television antenna to the 75-ohm input on the VCR, or the 75-ohm output of the VCR to the 300-ohm input of a television set, you have to use a transformer. Although these transformers might look different, there are only two general types. Internally they are all the same. The real difference is in the kinds of external connectors used. In all cases their job is to transform (change) the impedance of the signal.

   A standard television antenna will have screwhead type connectors for the spade lugs used with 300-ohm twin lead. A 300-to-75 transformer will have spade lugs to fit under these screws and a threaded connector on the other side to accept the F- type connector of a coaxial cable.

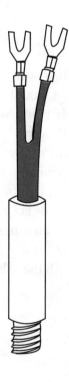

This same transformer can be used to connect a 75-ohm source (e.g., the VCR) to a 300-ohm input (such as a television set that has only this). In this case, 75-ohm coaxial cable is routed to the television set and is connected to the threaded side. The two spade lugs on the other side of the transformer connects to the television set. Used this way the transformer becomes a 75 to 300.

Another kind of 300-to-75 transformer is used when the wire is already 300 ohm but the VCR accepts only 75 ohm. This transformer has two screw heads for the spade lugs of the wire and a push-on F-type connector on the other side.

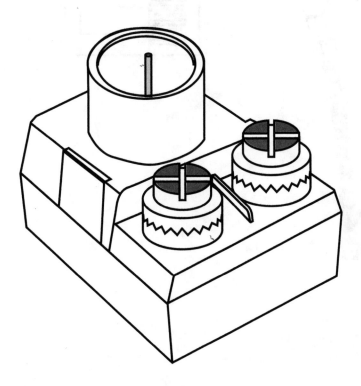

Some television sets have a built-in transformer, with an external switch (near the antenna input) that can be set to accept either 300-ohm or 75-ohm inputs. If your set has one, the kind of input and the switch setting must match.

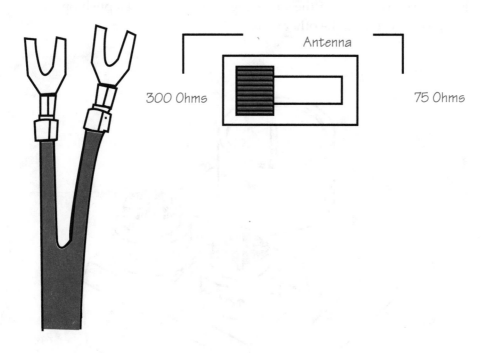

Step 9-5.
Connecting two VCRs.
If you intend to connect your VCR
to another VCR (for dubbing,
which is to copy the video
cassette in one machine to a
cassette in the other) or to
connect your VCR to a home
stereo system, you will need a
cable or cables with RCA-type
connectors. These are rarely
included with the VCR. One cable
will be needed for each signal you
wish to use. Only one is needed if
your goal is to connect a single audio
output to a stereo. Two are needed if the VCR
is capable of stereo, and two are needed for
dubbing (because you are using audio and
video outputs).

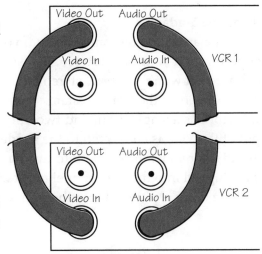

   The way to connect is
simply to plug the cable
into the needed output
of the VCR to the
needed input of the other
device. For example, to
dub you would connect the
Video Out of the first VCR to
the Video In of the second
VCR; and the Audio Out of
the first to the Audio In of
the other.

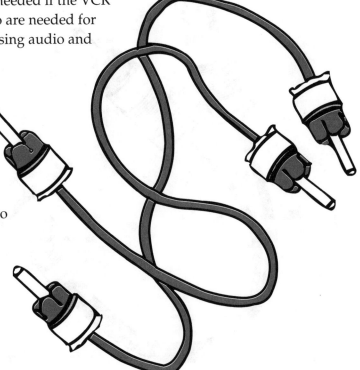

Step 9-6. Using the Y-adapter.

If your VCR is monaural and has only one audio output, you can still play the sound through your home stereo. By using a y-adaptor you can simulate stereo. It won't be "real" stereo but will be able to use the usually better speakers of the stereo system.

The y-adaptor has a single input and two outputs. The single audio output of the VCR is connected to the single input of the adaptor. The adaptor splits the signal into two parts. One of the connectors is plugged into the left channel input of the stereo amplifier and the other is plugged into the right channel input.

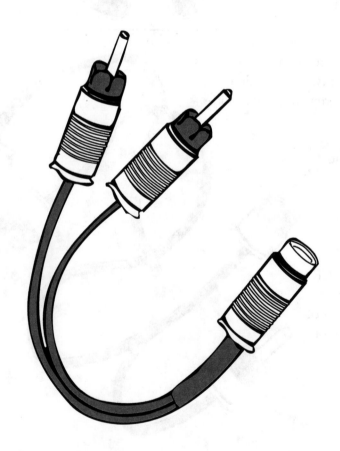

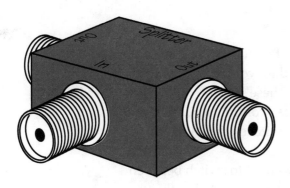

Step 9-7.
Using signal splitters.
To divide a signal coming in from one source, you can use a signal splitter. Signal splitters take one source and divide it into two or more outputs. You can use a signal splitter if you want to split an incoming cable signal to two VCRs. Buy a signal splitter at your local electronics store. Run the cables to each of the two VCRs.

Step 9-8.
Amplifying the signal.
If the signal is weak, or is split in too many directions, it might need to be amplified. A distribution amplifier does this. Connect the cable with the incoming signal to the input of the distribution amplifier. Connect the outputs to each of the units being driven, for instance two VCR and two television sets. You might also need a distribution amplifier if the unit being driven is a great distance away from the signal. Always put the amplifier near the source rather than the device.

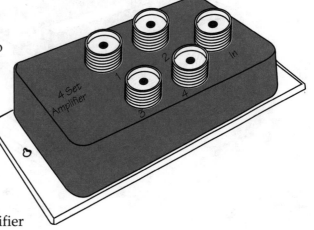

Step 9-9. Using an A-B switch.

If you want to switch one signal to two different outputs, you can use an A-B switch. The A-B switch is designed to work with the higher RF frequencies used in video and television. These switches allow you to switch between two sources. The switch can control two inputs and one output, or one input to two outputs. Switches that allow three or more outputs are also available.

You would use an A-B switch if you wanted to switch between cable and satellite signals. The two inputs are connected to the two inputs of the A-B switch. The output is connected to the device being driven (e.g., television or VCR).

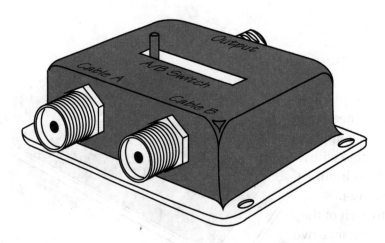

Step 9-10. Designing your system.

Now that you are aware of a few of the connectors, splitters, and amplifiers that are available, you can design the system you want. The easiest way to do this is to begin by drawing boxes to represent the different inputs and outputs of your system. This example shows two inputs (local antenna and cable service), and one VCR feeding two televisions. Notice the arrows that illustrate what the various parts do and the directions of the signals.

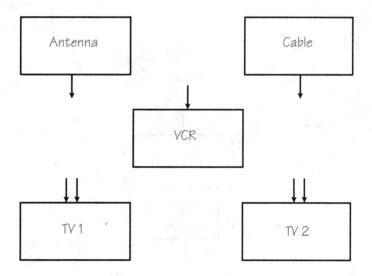

Step 9-11. Splitting the output signals.

Each of the two sources will be supplying a signal to the VCR and both televisions. This means that the single output of the antenna and cable will need to be split to provide three outputs. Likewise, the VCR is to provide two outputs—one to each television.

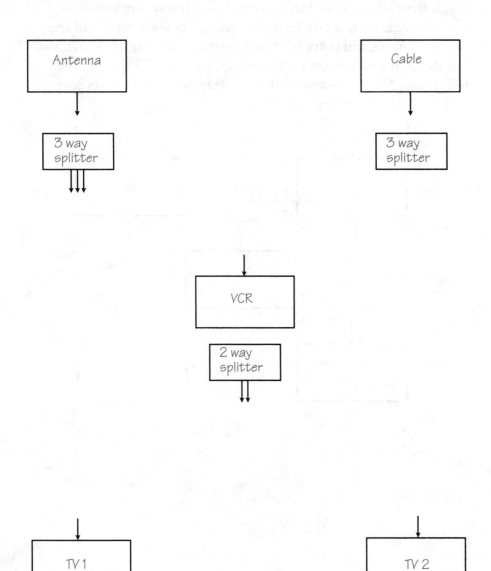

Step 9-12. Switching the input signals.

The VCR needs two inputs—one from the antenna and one from the cable. Because you want to take your choice of either the antenna or the cable service, you need to switch between the two. An A-B switch is used for this. Each of the televisions will now be receiving three signals (antenna, cable and VCR), and again switching is needed. This time you'll need an A-B-C switch (since there are three choices).

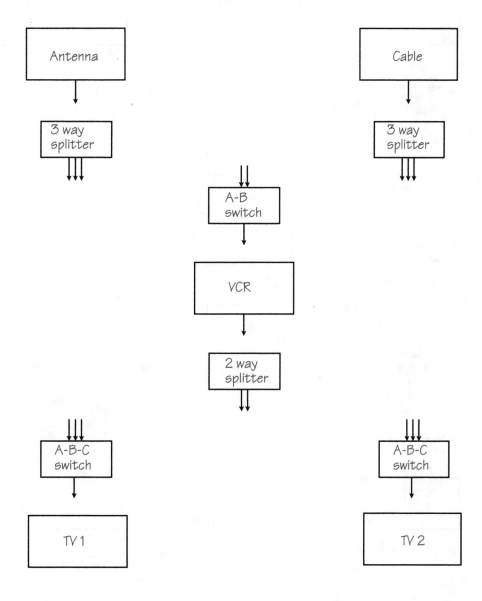

Step 9-13. Completing the design.

The only thing left is to label what comes from where, and what goes where. You can draw in the actual lines if you wish, but this isn't necessary. For the actual installation, it might help if the various cables are given labels (such as with a piece of tape).

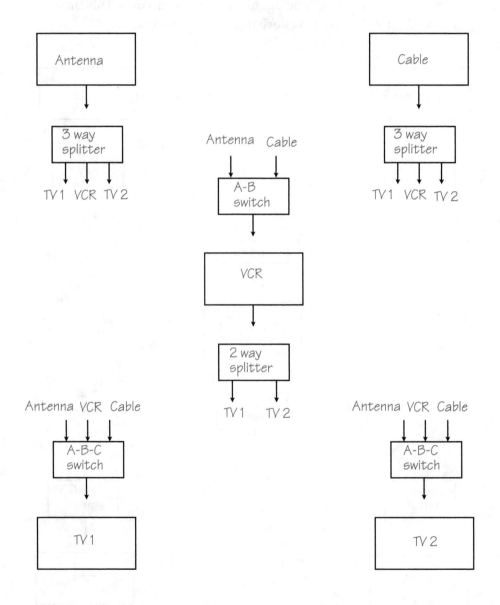

Once again, keep in mind that this is merely an example. The same technique can be used to design just about any system, regardless of how complicated it is. My own system has antenna and satellite TV inputs, with the VCR outputting to two televisions and to the home stereo system. Within that system all the components are also interconnected so that I can record or play the audio from the VCR, from the radio, from records or from the CD player. By flicking a switch here and there I can do two or three things at the same time. (For example, I can be recording a television program while watching a different channel, while the tape recorder is recording a radio program so I can listen to it later, while (if I wish) I listen to a new CD.

It sounds complicated, but the design was derived by the same techniques detailed above.

How to use a VOM

Some of the tests in this book require a volt-ohmmeter (VOM, also called a multimeter). A VOM is a handy and versatile piece of testing equipment, and it's easy to use.

With a VOM, you can test any batteries you use—including the one in your car. You also can test the wall outlets to see if they have power and if they have been wired correctly (and to make sure they are safe). You can test wires, cables, and connectors to see if they need to be replaced.

The VOM you buy doesn't have to be fancy or expensive. A model costing between $10 and $20 should suffice. Extreme accuracy isn't usually needed. When checking a wall outlet for power, you rarely need accuracy of greater than about 10 percent. Even the least expensive units are more accurate than that.

A VOM tests for ac voltage, dc voltage, and resistance. A number of ranges are available on the VOM for each test. The most common ac test checks the wall outlets in your home; your meter should have a setting for the 120-volt ac range. Common dc tests involve in or near typical battery outputs—namely 1.5, 3, 5-6, 9, and 12. Make sure your voltage setting is higher than the voltage you are testing, so you don't damage the VOM. For example, if you are expecting to read 120-volts ac, you would set the meter to the 150 setting. If you are uncertain of the value, start at the highest setting and work downwards.

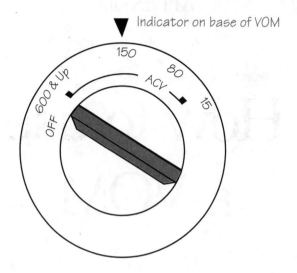

▼ Indicator on base of VOM

Step A-1. Setting up the meter.

For almost every test, plug the shorter end of the black lead into the "common" (–) jack on the meter. The shorter end of the red lead gets plugged into the + jack, which is often labeled as shown.

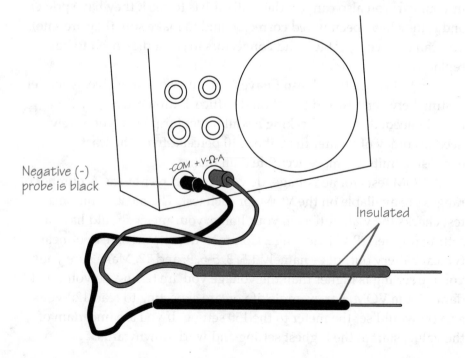

Negative (–) probe is black

-COM +V-Ω-A

Insulated

Step A-2. Setting the Ohms Adjust to zero.

To set the "Ohms Adjust," turn on the meter, set the dial to read resistance (any range), touch the two probes together, and turn the adjusting wheel until the reading is exactly zero.

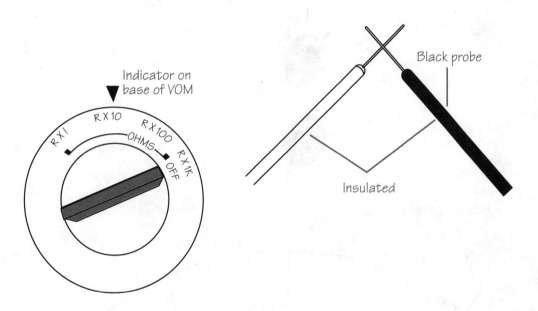

Indicator on base of VOM

Black probe

Insulated

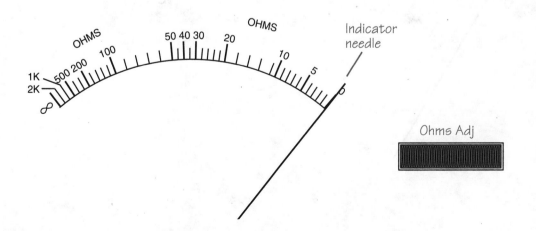

OHMS

Indicator needle

Ohms Adj

Step A-3. Setting the Zero Adjust.

If your meter has a "Zero Adjust," turn on the meter and adjust the control knob so that the meter reads exactly zero.

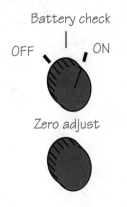

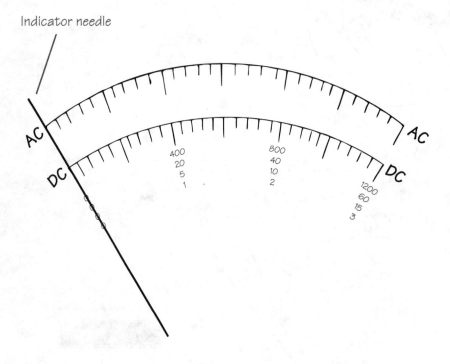

Step A-4. Reading dc volts.

To read dc volts (dc V or sometimes Vdc), set the meter to the proper range. Touch the black probe to the side with the negative (– or GND) label, and touch the red probe to the side with the positive (+) label.

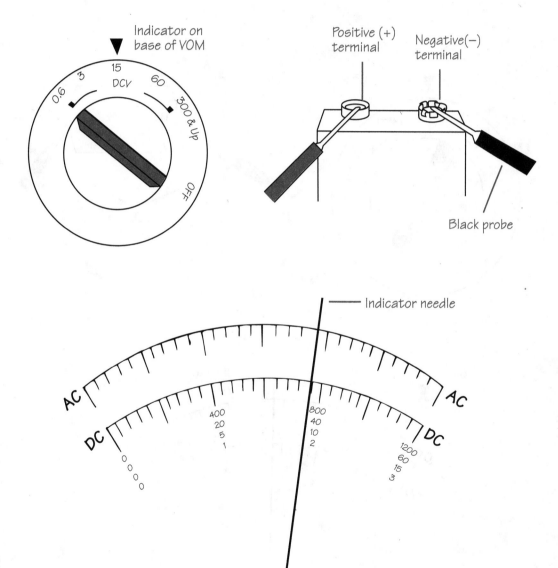

Step A-5. Testing a battery.

To test a battery, the positive terminal is usually round and is often labeled. Set the meter to read dc in the appropriate range. (AAA, AA, C, and D are all 1.5 volts. On most other batteries, the voltage is clearly labeled.)

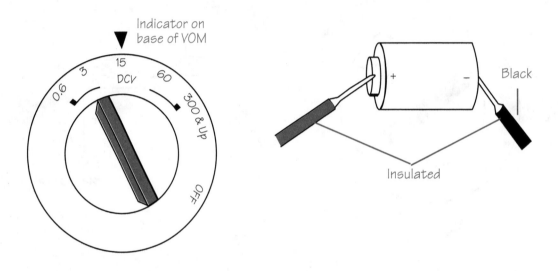

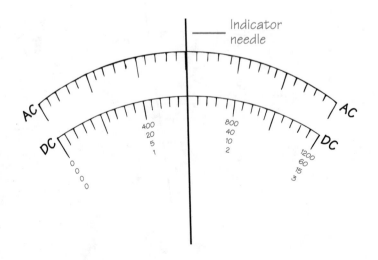

Step A-6. Testing for ac voltage.

When testing for ac voltage, the orientation of the probes is not important because the ac voltage is constantly changing from negative to positive and from positive to negative. Because ac is generally much more dangerous to you, it is essential that you hold the probes only by the insulated handles.

Black probe

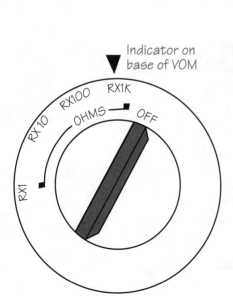

Indicator on base of VOM

Step A-7. Testing for continuity.

In this book, the most important resistance (ohms) test is for continuity, which simply means that the conductor being tested is continuous (not broken). Although the setting used is not important, it's best to use the highest setting (for resistance or ohms) on the dial.

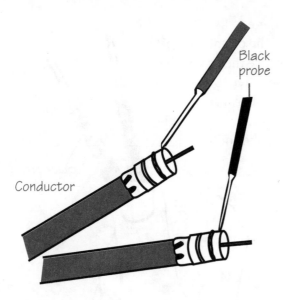

Black probe

Conductor

Step A-8. Checking for continuity.
Disconnect the wire or cable that is to be tested. Touch the probes to each end of the same conductor. It doesn't matter which probe touches which end. If the conductor is good, the needle should swing all the way to the right, giving a reading of zero ohms (or close to it). If there is a break in the wire, the reading will be close to infinity (full left scale and no needle movement).

Step A-9. Testing for a short.
To test for a short in a cable, touch one probe to the center conductor and the other to the outer ground. (Repeat this test using the center conductor at one end and the outer ground at the other.) If the cable is good, the reading should be infinity (full left scale and no needle movement). If the meter gives any reading (any needle movement), the two conductors are touching. If the meter swings all the way to zero ohms, the conductors are in direct contact, which is a short circuit; the cable is bad.

OHMS   50 40 30   20   OHMS

1K 500 200 100   10   5
2K   0
∞
———— Indicator needle

Step A-10.
Testing for dc on a circuit board.
When testing for dc on a circuit board, look for the label or GND. Touch the black probe to it and the red probe to the other dc point being tested. This will often be labeled something like +5 or +12.

Black probe

Glossary

ac Alternating current, such as the power from a household electrical outlet. The value is constantly changing in a sine wave.

antenna A device used to collect signals sent across the air from a television or radio transmitter.

aperture The opening of a lens; in video, this is more often called an iris, like the iris of the human eye.

audio Of or relating to sound. The audio portion of a video tape is recorded at the top of the tape if monaural. If stereo, it is recorded both at the top and in stripes along with the video portion.

a/v Audio/visual or audio/video. This is the combined signals of both the audio and visual portions.

azimuth To create the "striped" recording on the tape, the video head assembly is tilted at an angle. This angle is its azimuth.

battery An electrochemical device that generates electricity through a chemical reaction. A standard "cell" generates approximately 1½ volts dc, and a battery might consist of one or more cells. For example, a 9-volt battery has six cells inside it; a 12-volt battery has eight cells.

beta Video recording/playback system, invented by Sony. The cassette is physically smaller than a VHS cassette (but larger than a VHS-C, which is the "portable" version of VHS).

cable   A combination of two or more conductors in a single casing.

capstan   A vertical rotating shaft that drives tape at a constant speed. In a VCR the capstans are usually made of metal and often use a rubber roller.

cartridge   See cassette.

cassette   The plastic package containing tape.

CCD   Charge coupled device. The most common type of video pickup in home video cameras and camcorders. It has all but replaced the older tube-type pickups.

character generator   A device that generates letters, numbers, and other symbols. More sophisticated units can be set to change the styling and size of the characters.

chrominance   Color.

color temperature   Light has a variety of colors that your eye does not see. The light from an incandescent bulb, such as a standard household lamp, is more red than open sunlight; shade is more blue. A camera detects these differences and must be adjusted for them. This is done with the "white balance" control on the video camera. (See white balance.)

contrast   This is the balance or ratio between the brightest and darkest areas of a field. With "low contrast," the difference in tones is small; in "high contrast" the difference is large.

control track   Another term used to describe the sync track or cue track. The information recorded here controls the VCR and the television.

dc   Direct current, such as that in a battery.

distribution amplifier   A device that amplifies a signal and then distributes it to a number of other devices. For example, if you wish to use one VCR to supply a signal to five or six television sets, the signal might not be strong enough to drive them all. In such a case, a distribution amplifier is needed.

dockable   The camera and deck can be connected.

dropout   Loss of signal, usually caused by the magnetic coating of the tape flaking off, resulting in "sparkly" white specks on the television screen.

dub   To copy. Dubbing involves connecting one VCR to another, with one playing the tape and the other recording it as it plays. The original recording is called first generation, the first copy called second generation, a dub of this is third generation. Dubbing results in a decrease in quality from one generation to the next. Dubbing copyrighted tapes is illegal.

edit To modify in some way. In video this usually means removing unwanted scenes and juggling the order of scenes while making a dub. One of the easiest is "in-camera editing," which simply involves when you start and stop the camcorder.

erase head The electromagnet that erases the recorded signal on the tape so that a new signal can be recorded. In the VCR it usually comes before the video head assembly. (See also flying erase head.)

F-type The connector often used in home video equipment to carry composite signals. The center conductor of the cable becomes the prong that slides into the small hole of the connector on the VCR or television. The cable used with it is usually R-59.

flagging The apparent bending of the image at the top of the television screen. The most common cause is a worn or damaged sync track.

flying erase head An erase head mounted on the video head assembly, rather than separate from it. With the erase head being physically closer to the recording head, there are no glitches.

ghost A reflected or doubled image caused by two or more signals received simultaneously.

glitch Generally a recording error. A common glitch comes about when recording over an existing recording. Because of the distance between the erase head and the recording head, the tape is holding two different signals for a few seconds (until the erased tape gets to the recording heads). The result is a rainbowing effect, caused by the signals' interference.

head Any of the the electromagnets that record or playback. These include erase, video, and audio.

helical The video head assembly is tilted causing a spiral- like movement of the play-record heads. The recording is then placed on the tape in diagonal "stripes."

HQ High quality. One or more special circuits used in many VCRs to increase the image quality.

iris The opening (aperture) of the lens.

loading platform Platform that holds the tape flat during loading.

luminance Brightness.

lux The metric measurement of light quantity. It is the amount of light reflected from a surface of one square meter by a standard candle one meter away. It is approximately equal to 11 foot-candles.

macro A lens feature that allows the lens to be focused while very close to an object.

monitor The television set. In more technical circles, a monitor is a special kind of television that exhibits a better image than does the usual home television set.

nicad nickel-cadmium. A type of rechargable battery.

ohm The unit of resistance to the flow of electrical current.

oxide The result when oxygen combines chemically with another element. In recording, various oxides are often used as the magnetic particles on the tape.

phono plug Also called an RCA-type plug. Characterized by a fairly large center prong and "wings" around the outside, this plug is often used to carry the separate audio and video signals from one VCR to another for dubbing. Another common use is for carrying the audio signals of a stereo system.

pickup The device inside the video camera that converts the light into electricity, which in turn is used to create the recording.

pixel Picture element; tiny spots that make up the overall picture.

RCA plug See phono plug.

RF Radio frequency; the band of frequencies used for radio and television signals.

roller A "wheel" usually made of rubber (sometimes plastic) and used in the tape path to keep the tape moving correctly.

SVHS Super-VHS; a relatively new system designed to increase the quality of the recording and playback over standard VHS.

S/N Signal-to-noise ratio. All recordings, amplifiers, etc. have some unwanted signals inherent. The more noise there is compared to the wanted signal, the poorer quality you'll get.

sync Short for synchronization; keeping all the signals of the VCR and television "in step." The pulses that control this are recorded on the sync track of the tape, which is a linear recording at the bottom of the tape.

tracking Adjustment of the video head assembly that "times" the movement of the playback/recording heads across the tape.

UHF   Ultrahigh frequency; a band of frequencies above VHF making up the television channels 15 and above.

VCR   Video cassette recorder; a VCR is always a VTR (see VTR), but a VTR isn't necessarily a VCR.

VHF   The band of television frequencies making up channels 2 through 14. There is no channel 1.

VHS   Recording/playback system invented by JVC. VHS-C uses the same formatting with a smaller cassette.

VOM   Volt-ohmmeter, also called a multimeter because it can measure voltage, resistance (measured in ohms) and often current (measured in amps).

VTR   Video tape recorder. See also VCR.

white balance   The control on a video camera that allows the operator to adjust it for the kind and color of the ambient light. (See also color temperature.)

zoom   A lens feature that allows the user to get a closer picture of the subject without moving physically. A 10:1 ratio allows you to videograph a subject at 100 feet as though it was only 10 feet away.

Index